CAMBRIDGE LIBRARY COLLECTION

Books of enduring scholarly value

Life Sciences

Until the nineteenth century, the various subjects now known as the life sciences were regarded either as arcane studies which had little impact on ordinary daily life, or as a genteel hobby for the leisured classes. The increasing academic rigour and systematisation brought to the study of botany, zoology and other disciplines, and their adoption in university curricula, are reflected in the books reissued in this series.

The Natural History of Birds

Georges-Louis Leclerc, Comte de Buffon (1707–88) was a French mathematician who was considered one of the leading naturalists of the Enlightenment. An acquaintance of Voltaire and other intellectuals, he work as Keeper at the Jardin du Roi from 1739, and this inspired him to research and publish a vast encyclopaedia and survey of natural history, the ground-breaking *Histoire Naturelle*, which he published in forty-four volumes between 1749 and 1804. These volumes, first published between 1770 and 1783 and translated into English in 1793, contain Buffon's survey and descriptions of birds from the *Histoire Naturelle*. Based on recorded observations of birds both in France and in other countries, these volumes provide detailed descriptions of various bird species, their habitats and behaviours and were the first publications to present a comprehensive account of eighteenth-century ornithology. Volume 1 covers birds of prey and flightless birds.

Cambridge University Press has long been a pioneer in the reissuing of out-of-print titles from its own backlist, producing digital reprints of books that are still sought after by scholars and students but could not be reprinted economically using traditional technology. The Cambridge Library Collection extends this activity to a wider range of books which are still of importance to researchers and professionals, either for the source material they contain, or as landmarks in the history of their academic discipline.

Drawing from the world-renowned collections in the Cambridge University Library, and guided by the advice of experts in each subject area, Cambridge University Press is using state-of-the-art scanning machines in its own Printing House to capture the content of each book selected for inclusion. The files are processed to give a consistently clear, crisp image, and the books finished to the high quality standard for which the Press is recognised around the world. The latest print-on-demand technology ensures that the books will remain available indefinitely, and that orders for single or multiple copies can quickly be supplied.

The Cambridge Library Collection will bring back to life books of enduring scholarly value (including out-of-copyright works originally issued by other publishers) across a wide range of disciplines in the humanities and social sciences and in science and technology.

The Natural History of Birds

From the French of the Count de Buffon

VOLUME 1

COMTE DE BUFFON
WILLIAM SMELLIE

CAMBRIDGE UNIVERSITY PRESS

Cambridge, New York, Melbourne, Madrid, Cape Town, Singapore,
São Paolo, Delhi, Dubai, Tokyo, Mexico City

Published in the United States of America by Cambridge University Press, New York

www.cambridge.org
Information on this title: www.cambridge.org/9781108022989

© in this compilation Cambridge University Press 2010

This edition first published 1793
This digitally printed version 2010

ISBN 978-1-108-02298-9 Paperback

THE

NATURAL HISTORY

OF

B I R D S.

FROM THE FRENCH OF THE

COUNT DE BUFFON.

———

ILLUSTRATED WITH ENGRAVINGS;

AND A

PREFACE, NOTES, AND ADDITIONS,

BY THE TRANSLATOR.

———

IN NINE VOLUMES.

VOL. I.

LONDON:

PRINTED FOR A. STRAHAN, AND T. CADELL IN THE STRAND;
AND J. MURRAY, N° 32, FLEET-STREET.

MDCCXCIII.

PREFACE,

BY THE TRANSLATOR.

———————

Few writers have been more juſtly admired for originality, and grandeur of conception, than the celebrated Comte de Buffon. It was his lively eloquence that firſt reſcued Natural Hiſtory from barbariſm, and rendered it an engaging and popular ſtudy. With concern and indignation he beheld the faireſt of all the ſciences cramped by artificial ſyſtems, encumbered by a coarſe and obſcure jargon, and disfigured by credulity and ignorance. He was determined to reſtore and decorate the fabric. Royal munificence happily ſeconded his views ; and he was entruſted with the direction of the fineſt cabinet in Europe. His lofty genius burſt from the ſhackles of method ; he caught with ardour the varied magnificence of Nature's plan ; and, with a maſterly pencil, dipt in rich and glowing colours, he traced the animated picture. His

elegant

elegant and spirited diction adorns whatever subject he treats ; his various and extensive learn-
ing at once pleases and instructs. His graceful
turn of sentiments engages our affections ; the
sublimity of his descriptions commands our admiration ; and if the exuberance of his fancy
has sometimes laid him open to censure, we are
disposed to overlook his errors for the brilliancy
of his composition.

His Theory of the Earth was first published
in 1744 ; his History of Man soon followed ;
but that of Quadrupeds was not completed till
1767. The History of Birds was next to be
undertaken, a task attended with peculiar difficulties. The species of Birds are at least ten
times more numerous than those of Quadrupeds, and are subject to endless varieties. Their
mode of life exposes them to the immediate influence of the seasons ; in a large proportion of them the migrations to remote climates produce important alterations on their
external appearance ; and their hot temperament sometimes perverts their instincts, and
gives birth to unnatural progeny that serve to
increase the confusion. The dispositions and
œconomy of Birds are in a great measure removed from observation ; and our knowledge,
with regard to them, is necessarily scanty and
imperfect. But M. de Buffon was not to be
deterred by the difficulty and extent of the
undertaking. The correspondents of the king's
                                              cabinet

cabinet continued to tranfmit numerous communications, and fpecimens from all parts of
the world. Above eighty artifts were under
the direction of the younger M. Daubenton,
employed five years in the drawing, engraving,
and colouring, of upwards of a thoufand Birds.
But the commencement of the work which
thefe were intended to illuftrate was delayed
two years, by reafon of a fevere and tedious
indifpofition, which during that fpace afflicted
the excellent Naturalift. And after he had re-
covered his health, he reflected that at his advanced period of life he could not reafonably
expect to be able to accomplifh the Hiftory of
Birds, and alfo that of Minerals, in which he
had already made fome advances. He judged
it expedient therefore to have recourfe to the
affiftance of his friends ; and he was peculiarly
fortunate in the choice of the learned and elo-
quent M. Gueneau de Montbeillard, who cheer-
fully undertook the laborious tafk, and compofed the greateft part of the two firft volumes of
the Hiftory of Birds, which appeared in 1771,
under the name however of M. de Buffon. In
his complexion of thought and mode of expreffion, M. de Montbeillard followed fo clofely
his illuftrious affociate, that the Public could
not perceive any change. It was now proper
to throw off the mafk ; and in the publication
of the four fubfequent volumes, each author
prefixed his name to his own articles. The

third

third volume was nearly printed when new affiftance was received from the communications
of James Bruce, Efq. of Kinnaird. That accomplifhed and adventurous traveller in his
return from Abyffinia paffed fome days with
M. de Buffon at Paris. The Count was filled
with admiration on feeing the numerous and
elegant drawings which Mr. Bruce had made
of natural objects; and on feveral occafions he
mentions the explorer of the fource of the Nile
in terms the moft flattering and refpectful. After the publication of the fixth volume in 1781,
M. de Montbeillard was defirous of devoting the
whole of his leifure in compofing the Hiftory of
Infects, which had become his favourite ftudy.
The three remaining volumes were therefore
written by M. de Buffon himfelf; though he
acknowledges that the Abbé Bexon had collected the nomenclature, formed moft of the defcriptions, and communicated feveral important
hints. The work was completed in 1783; and
as only a few copies of the Illumined Plates
were on fale, and thefe extremely coftly, a fmall
fet of engravings were made to accommodate
ordinary purchafers. M. de Buffon had about
the fame time finifhed his Hiftory of Minerals.
He now entertained views of compofing the
Hiftory of Vegetables, in which delightful fubject his ingenuity, his tafte, and his erudition,
eminently qualified him to fhine; but unfortu-
nately for the Public the project was defeated

by

by the death of that great man on the 16th of
April 1788.

To expatiate on the advantages arifing from
an acquaintance with Natural Hiftory might be
deemed unneceffary. It affords an elegant and
rational fpecies of entertainment; and as it requires no previous courfe of ftudy, it feems admirably fitted to captivate the minds of youth,
and to fix their attention. It difpels many early
prejudices, raifes and warms their opening fancy,
enlarges the circle of their ideas, and leads by
eafy and flowery fteps to the purfuit of the ab-
ftrufer fciences. It conveys ufeful and interefft-
ing information refpecting the fituation of the various countries, their climate, their productions,
and the manners and œconomy of the inhabitants. But above all, the contemplation of that
order and defign, fo confpicuous in the works of
Nature, allays the ftormy paffions, elevates the
foul to virtue and happinefs, and exhibits the
moft enchanting profpects of that wifdom and
power which upholds and conducts the univerfe.

Books of Natural Hiftory feem, more than
any others, to require tranflation. They muft
unavoidably abound with uncommon words and
phrafes, which frequently create difficulties even
to proficients in the language; the vivacity of
the impreffion is at any rate weakened; and the
reading, inftead of fafcinating by the pleafure
which it is calculated to afford, degenerates per-

A 4

haps

haps into an irkfome tafk. The names of quadrupeds, of birds, of fifhes, of infects, and reptiles, of plants, and of minerals, are befides hardly ever explained accurately in dictionaries, and are
frequently omitted altogether. There are many
perfons who might be deterred by the expence
from purchafing the original, or who, from their
fituation and circumftances in life, have not had
leifure or opportunities of acquiring a compe-
tent knowledge of the language in which it is
compofed. To accommodate this numerous clafs
of readers, to increafe the circulation of ufeful
and popular works, is the chief object of tranflation. A diffufion of tafte and information
forms the diftinguifhing feature of our own times.
Men of a gloomy or fplenetic temper may de-
claim againft the frivoloufnefs of the age : to
decry the prefent and extol the paft, is indeed
an inveterate, an incurable malady. Other
periods have produced great and fhining characters, who foared above the prejudices and
narrow views of their contemporaries. But a
liberality of fentiment, unknown to our rude
forefathers, now generally prevails ; the fweeteft
of all the virtues, and that which contributes
the moft to alleviate the ills and heighten the
joys of life, humanity and fellow-feeling, has fhed
its lovely influence on all ranks ; and never did
the fun behold fuch a large portion of mankind
fo enlightened, fo refpectable, and fo happy.

The

The great expence attending the publication
of an extensive work, adorned with numerous
plates, has long prevented Buffon's Natural History from appearing in an English dress. It is
only a few years since a translation of the first part
was given by Mr. Smellie of Edinburgh; and
the favourable reception which this has met
with, attests sufficiently its merit. But that gen-
tleman has not chosen to complete the task. The
History of Minerals indeed, though replete with
curious and often solid information, is addressed
to a narrow circle of readers. But the History
of Birds possesses every quality that could re-
commend it to the public: it exhibits a clear and
comprehensive view of the knowledge acquired
in Ornithology, scattered through a multiplicity
of volumes and in various languages; it discusses
and elucidates, with critical accuracy, the numerous controverted points; it reduces the whole
to simplicity, order, and elegance; and, by large
additions of valuable matter, it greatly extends
the bounds of the science.

In translating this work, I have studied to
transfuse the spirit of the author into our
language. I was aware of the tendency to
adopt foreign idioms, and I was solicitous to
avoid that censure. How far I have suc-
ceeded, the public will judge. Zoological descriptions aim not only at perspicuity, but require the most minute accuracy; in such parts,
therefore, where the subject assumes a loftier
tone,

tone I have ftuck clofe to the original. I have
endeavoured to obferve a correfponding elevation
of ftyle. There are fome fprightly turns in the
French which the mafculine character of our language will not admit ; but thefe inferior beauties
are amply compenfated by the ftrength and dig-
nity of its expreffion. The philofophy likewife
of that ingenious people has a certain diffufe
fuperficial caft, not altogether fuited to the manly
fenfe of the Britifh nation. The tranflator fhould
have a regard to the tafte of his countrymen whom
he addreffes ; and, on proper occafions, he may,
with advantage, be permitted to abridge and
condenfe.

I have difcovered in the text a few inaccuracies which I have taken the liberty to correct.
A few notes which I have fubjoined, will ferve
to elucidate the paffages. I have confulted the
lateft authors who have either written exprefsly
on Ornithology, or who have occafionally
handled the fubject ; and the additions which
I have thereby been enabled to make, will, I truft,
prove not unacceptable. I have beftowed particular attention to the nomenclature, which it
is the principal aim of fyftems to fix and afcertain. Thefe productions will, no doubt, rank
very low in the eftimation of the philofopher ;
yet they muft ftill be regarded as ufeful helps towards the ftudy of Natural Hiftory. It was the
want of them that fo often occafions fuch obfcurity and uncertainty in the writings of the
ancient

ancient naturalifts. If to difcover the name of an
animal or a vegetable, we were obliged to fearch
over and compare a whole feries of defcriptions,
the fatigue would be intolerable. No perfon
objects to a dictionary, becaufe the words follow
alphabetically, and not according to their gradation of meaning. If by means of arrangement, how artificial foever, we can, from a few
obvious characters, refer an object fucceffively to
its order, its genus, and its fpecies, we fhall trace
out its name, and thence learn its properties with
eafe and pleafure: and even though contiguous
divifions always run into one another, the number of poffible trials is at any rate much limited,
and the labour of the inveftigation abridged.
To complete Natural Hiftory requires the union
of Buffon and Linnæus. With this view therefore, I have given an abftract of the Linnæan
claffification of Birds from the laft edition of his
Syftema Naturæ, by Gmelin, in 1788; and to
each article of the work I have joined his names
and fynonyms, with a tranflation of the fpecific
character. Moft of the other additions I owe
to Mr. Latham, and particularly to Mr. Pennant:
I fhould be ungrateful did I not acknowledge the
affiftance which I have received from the various and entertaining works of this amiable naturalift.

But notwithftanding the pains which I have
beftowed to render this work complete, I deliver
it to the public with the anxiety that naturally
accom-

accompanies a firſt attempt. It is compoſed at an early period of life, and in the retirement of the country. Some inaccuracies and blemiſhes may have eluded my attention. Motives of prudence will determine me to withhold my name: for hard is the lot of the tranſlator; his humble toil is commonly beheld with diſdain; and the utmoſt he can expect is to eſcape cenſure. This ſupercilious treatment has already occaſioned pernicious effects. Men of ſuperior talents have generally deſerted a path that leads neither to honour nor emolument. Hence the purity of our language has been violated by an inundation of vicious and foreign idioms, and tranſlations have often been written, that really merit contempt. Should the public alſo frown upon *my* labours, I ſhall at leaſt conſole myſelf with the hope, that the experience of maturer years may correct my errors. But if it will deign to receive this work with indulgence (this is all that I intreat), the approbation will animate my exertions and heighten my enjoyments.

CONTENTS

OF THE

FIRST VOLUME,

The

CONTENTS.

The

CONTENTS.

CONTENTS.

O N

ON THE

NATURE

OF

BIRDS.

THE word Nature has in all languages two very different acceptations. It denotes either that Being, to the operation of which we ufually afcribe the chain of effects that conftitute the phænomena of the univerfe; or it fignifies the aggregate of the qualities implanted in man, or in the various quadrupeds, and birds, &c. It is *active* nature that, ftamping their peculiar characters, thus forms *paffive* nature; whence are derived the *inftincts* of animals, their *habits*, and their *faculties*. We have in a former work treated of the nature of Man and the Quadrupeds; that of Birds now demands our attention: and though the fubject is, in many refpects, more obfcure, we fhall endeavour to felect the difcriminating features, and to place them in the proper point of view.

Per-

Perception, or rather the faculty of feeling; inftinct, which refults from it; and talent, which confifts in the habitual exercife of the natural powers; are widely diftinguifhed in different beings. Thefe intimate qualities depend upon organization in general, and efpecially upon that of the fenfes: they are not only proportioned to the degree of the perfection of thefe; they have alfo a relation to the order of fuperiority that is eftablifhed. In man, for inftance, the fenfe of touch is more exquifite than in all other animals; in thefe, on the contrary, fmell is more perfect than in man: for touch is the foundation of knowledge, and fmell is only the fource of perception. But, as few perfons diftinguifh nicely the fhades that difcriminate between ideas and fenfations, knowledge and perception, reafon and inftinct, we fhall fet afide what are termed *ratiocination, difcernment,* and *judgment;* and we fhall only confider the different combinations of fimple perception, and endeavour to inveftigate the caufes of that diverfity of inftinct, which, though infinitely varied in the immenfe number of fpecies, feems more conftant, more uniform, and more regular, and lefs fubject to caprice and error, than reafon in the fingle fpecies which boafts the poffeffion of it.

In comparing the fenfes, which are the primary powers that readily excite and impel the inftinct in all animals, we find that of fight to
be

be more extended, more acute, more accurate, and more diftinct in the birds in general, than in the quadrupeds : I fay in general, for there are fome birds, fuch as the owls, that have lefs clear vifion than the quadrupeds ; but this, in fact, refults from the exceffive fenfibility of the eye, which, though it cannot fupport the glare of noon-day, diftinguifhes nicely objects in the glimmering of the evening. In all birds the organ of fight is furnifhed with two membranes, an external and internal, additional to thofe which occur in man : the former *, or external membrane, is placed in the large angle of the eye, and is a fecond and more tranfparent eyelid, whofe motions too are directed at pleafure, and whofe ufe is to clear and polifh the cornea : it ferves alfo to temper the excefs of light, and confequently to adjuft the quantity admitted, to the extreme delicacy of the organ : the † other

is

---

* This internal eye-lid *(membrana nictitans)* occurs in feveral quadrupeds ; but in moft of them it is not moveable as in birds.

† " In the eyes of a turkey cock, the optic nerve, which was fituated very near the fide, after perforating the fclerotic and choroid coats, fpread into a round fpace, from the circumference of which a number of black filaments were fent off to form by their union a membrane which is found *in all birds.*"——" In the eyes of the oftrich, the optic nerve, after perforating the fclerotic and choroid coats, was dilated into a fort of funnel of a fimilar fubftance : this funnel is not commonly round in birds, where we have almoft always found the extremity of the optic nerve flattened and compreffed within the eye : from this funnel a folded membrane took its origin, forming a fort of purfe that drew to a point. This purfe, which was fix lines broad at the bafe, where it grew out of the op-

tic

is fituated at the bottom of the eye, and appears
to be an expanfion of the optic nerve, which,
receiving more immediately the impreffions of
the rays, muft be much more fenfible than in
other animals ; and hence the fight is in birds
vaftly more perfect, and embraces a wider range.
A fparrow-hawk, while he hovers in the air,
efpies a lark fitting on a clod, though at twenty
times the diftance at which a man or dog could
perceive it.   A kite which foars to fo amazing
a height as totally to vanifh from our fight, yet
diftinguifhes the fmall lizards, field-mice, birds,
&c. and from this lofty ftation he felects what
he deftines to be victims of his rapine. But this
prodigious extent of vifion is accompanied likewife with an equal accuracy and clearnefs ; for
the eye can dilate or contract, can be fhaded or
uncovered, depreffed or made protuberant, and
thus it will readily affume the precife form fuited
to the quantity of light and the diftance of the
object.

Sight has a reference alfo to motion and
fpace ; and, if birds trace the moft rapid courfe,
we might expect them to poffefs in a fuperior
degree that fenfe which is proper to guide and
direct their flight. If Nature, while fhe endowed them with great agility and vaft mufcular

tic nerve, was black, and feemed imbued, and quite penetrated by
that colour, which on the choroid is only fpread, and may be rub-
bed off with the fingers.'' *Memores pour fervir à l'Hift. des Anim.*

ftrength,

ftrength, had formed them fhort-fighted, their latent powers would have availed them no-thing; and the danger of dafhing againft every intervening obftacle would have repreffed or ex-tinguifhed their ardour. Indeed, we may con-fider the celerity with which an animal moves, as the juft indication of the perfection of its vi-fion. A bird, for inftance, that fhoots fwiftly through the air, muft undoubtedly fee better than one which flowly defcribes a waving tract. Among the quadrupeds too, the *floths* have their eyes enveloped, and their fight is limited.

The idea of motion, and all the other ideas which accompany or flow from it, fuch as thofe of relative velocities, of the extent of country, of the proportional height of eminences, and of the various inequalities that prevail on the furface, are, therefore, more precife in birds, and occupy a larger fhare of their conceptions than in qua-drupeds. Nature would feem to have pointed out this fuperiority of vifion by the more con-fpicuous and more elaborate ftructure of its or-gan; for in birds the eye is larger in proportion to the bulk of the head than in quadrupeds * ;

* " The ball of the eye in a female eagle was, at its greateft width, an inch and half in diameter; that of the male was three lines lefs." *Mem. pour fervir à l' Hift. des Animaux.*

The ball of the ibis' eye was fix lines in diameter.

The eye of the ftork four times larger. *Idem.*

The ball of the caffowary's eye was four times larger than its *cornea*, being an inch and half in diameter, though the cornea was only three lines. *Idem.*

it

it is alſo more delicate and more finely faſhion-
ed, and the impreſſions which it receives muſt
excite more vivid ideas.

Another cauſe of the difference between the
inſtinﬅs of birds and of quadrupeds, is the nature of the element in which they live.   The
birds know better than man, perhaps, all the degrees of reſiſtance of the air, its temperature at
different heights, its relative denſity, &c.   They
foreſee more than us, they indicate better than
our barometers or thermometers, the changes
which happen in that voluble fluid.   Often have
they ſtruggled againſt the violence of the wind,
and oftener have they borrowed its aid.   The
eagle, ſoaring above the clouds *, can quickly
eſcape from the ſcene of the ſtorm to the region
of calm, and there enjoy a ſerene ſky and a
bright ſun, while the other animals below are
involved in darkneſs, and expoſed to all the
fury of the tempeſt.   In twenty-four hours it
can change its climate, and ſailing over the different countries, it will form a piﬅure which
exceeds the powers of our imagination.   Our
bird's-eye views, of which the accurate execution
is ſo tedious and ſo difficult, give very imperfeﬅ

---

* It can be proved that the eagle, and other birds of lofty flight,
can riſe perpendicularly above the clouds; for they frequently mount
entirely out of our ſight.   But in day-light an objeﬅ ceaſes to be
viſible when it exceeds 3,436 times its diameter; if, therefore, the
extent of the bird be five feet, it will be ſeen at the height of 17,180
feet, or above three miles. ,

notions

notions of the relative inequality of the furfaces which they reprefent. But birds can chufe the proper ftations, can fucceffively traverfe the field in all directions, and with one glance comprehend the whole. The quadruped knows only the fpot where it feeds; its valley, its mountain, or its plain: it has no conception of the expanfe of furface, no idea of immenfe diftances, and no defire to puſh forward its excurfions. Hence remote journies and migrations are as rare among the quadrupeds as they are frequent among the birds. It is this defire, founded on their acquaintance with foreign countries, on the confcioufnefs of their expeditious courfe, and on their forefight of the changes that will happen in the atmofphere and of the revolution of feafons, that prompt them to retire together, and by common confent. When their food begins to grow fcarce, when, as the cold or the heat incommodes them, they refolve on their retreat, the parents collect their young, and the different families affemble and communicate their views to the unexperienced; and the whole body, ftrengthened by their numbers, and actuated by the fame common motives, wing their journey to fome diftant land.

This propenfity to migration, which recurs every fpring and autumn, is a fort of violent longing, which, even in captive birds, burfts out in fymptoms of reftlefs and uneafy fenfations. We ſhall, at the article of the Ouail, give a detail

of

of obfervations on this fubject; from which it
will appear, that this propenfity is one of their
moft powerful inftincts; and that, though they
ufually remain tranquil in their prifon, they
make every exertion at thofe periods to regain
their liberty, and join their companions.——But
the circumftances which attend migration vary
in different birds; and, before we enter into the
full difcuffion which that fubject merits, we
fhall purfue our inveftigation of the caufes that
form and modify their inftincts.

Man is eminently fuperior to all the animals
in the fenfe of touch, perhaps too in that of
tafte; but he is inferior to moft of them in the
other three fenfes. When we compare the animals with each other, we foon perceive that
fmell in general is more acute among the quadrupeds than among the birds: for though we
fpeak of the fcent of the crow, of the vulture, &c.
it undoubtedly obtains in a much lower degree;
and we might be convinced of this by merely
examining the ftructure of the organ. In moft
of the winged tribes, the external noftrils are
wanting, and the effluvia, which excite the fenfation, have accefs only to the duct leading from
the palate*: and even in thofe where the organ is difclofed, the nerves, which take their
origin from it, are far from being fo numerous,
fo large, or fo expanded, as in the quadruped.

* Hift. de l'Acad des Sciences, tome i. p. 430.

We

We may therefore regard touch in man, fmell in the quadruped, fight in the bird, as the three moft perfect fenfes, and which influence the general character.

Next to fight, the moft perfect of the fenfes in birds is hearing, which is even fuperior to that of the quadrupeds. We perceive with what facility they retain and repeat tones, fuc-ceffions of notes, and even difcourfe; we delight to liften to their unwearied fongs, to the inceffant warbling of their happy loves. Their ear and throat are more ductile and more powerful than in other animals. Moft of the quadrupeds are habitually filent; and their voice, which is feldom heard, is almoft always harfh and difagreeable. In birds it is fweet, pleafant, and melodious. There are fome fpecies, indeed, in which the notes feem unfupportable, efpeci-ally if compared with thofe of others; but thefe are few in number, and comprehend the large kinds, which Nature, beftowing on them hoarfe loud cries, fuited to their bulk, would incline to treat like quadrupeds. A peacock, which is not the hundredth part of the fize of an ox, may be heard farther; the nightingale could fill a wider fpace with its mufic than the human voice: this prodigious extent, and the great powers of their organs of found, depend entirely on the ftructure; but that their fong fhould be continued and fupported, refults folely from their internal

internal emotions. Thefe two circumftances ought to be confidered feparately.

The pectoral mufcles are more flefhy and much ftronger in birds than in man or the quadrupeds, and their action is immenfely greater. Their wings are broad and light, compofed of thin hollow bones, and connected by powerful tendons. The eafe with which birds fly, the celerity of their courfe, and even their power of directing it upwards or downwards, depend on the proportion of the impelling furface to the mafs of the body. When they are ponderous, and the wings and tail at the fame time fhort, like the buftard, the caffowary, or the oftrich, they can hardly rife from the ground.

The windpipe is wider and ftronger in birds than in quadrupeds, and ufually terminates below in a large cavity that augments the found. The lungs too have greater extent, and fend off many appendices which form air-bags, that at once affift the motion, by rendering the body fpecifically lighter, and give additional force to the voice. A little production of the cartilage of the *trachea* in the howling baboon *, which is a quadruped of a middle fize only, and of the ordinary ftructure, has enabled it to fcream almoft without intermiffion, and fo loud, as to be heard at more than a league's diftance : but in birds,

* Simia—Bcelzebut. *Linn.*

the

the formation of the thorax, of the lungs, and of all the organs connected with thefe, feems exprefsly calculated to give force and duration to their utterance; and the effect muft be pro-portionally greater *.

There is another circumftance which evinces that birds have a prodigious power of voice: the cries of many fpecies are uttered in the higher regions of the atmofphere, where the rarity of the medium muft confequently weaken the effect. That the rarefaction of the air di-minifhes founds is well afcertained from pneu-matical experiments; and I can add, from my own obfervation, that, even in the open air, a fenfible difference in this refpect may be per-ceived. I have often fpent whole days in the forefts, where I was obliged to liften clofely to

* In moft water-fowls, which have a very ftrong voice, the *trachea* reverberates the found; for the *glottis* is placed below it, and not above it, as in man. *Coll. Acad. Part. Fr. tome* i. 496.— The fame is the cafe in the cock. *Hift. de l'Acad. tome* ii. 7. In birds, efpecially ducks and other water-fowls, the organs of voice confift of an *internal larynx* placed where the *trachea arteria* parts; of two membranous pipes which communicate below with the two firft branches of the *trachea*; of many femilunar membranes, dif-pofed one above another in the principal branches of the flefhy lungs, and which, occupying only one half of their cavity, allow a free exit to the air; of other membranes placed in various po-fitions, both in the middle and in the lower part of the *trachea*; and laftly, of a membrane, of more or lefs folidity, fituated al-moft tranfverfely between the two branches of the *lunula*, which terminates a cavity that conftantly occurs in the upper and inter-nal part of the breaft. *Mem. de l'Acad. des Sciences, anné* 1753.

the

the diftant cries of the dogs, or fhouts of the
hunters; I uniformly found that the fame noifes
were much lefs audible during the heat of the
day, between ten and four o'clock, than in the
evening, and particularly in the night, whofe
ftillnefs would make hardly any alteration, fince
in thefe fequeftered fcenes there is nothing to
difturb the harmony but the flight buzz of infects and the chirping of fome birds. I have
obferved a fimilar difference between the frofty
days in winter and the heats of fummer. This
can be imputed only to the variation in the
denfity of the air. Indeed, the difference feems
to be fo great, that I have often been unable to
diftinguifh in mid-day, at the diftance of fix
hundred paces, the fame voice which I could, at
fix o'clock in the morning or evening, hear at
that of twelve or fifteen hundred paces.—A bird
may rife at leaft to the height of feventeen
thoufand feet, for it is there juft vifible. A
flock of feveral hundred ftorks, geefe, or ducks,
muft mount ftill higher, fince, notwithftanding
the fpace which they occupy, they foar almoft
out of fight. If the cry of birds therefore may
be heard from an altitude of above a league,
we may reckon it at leaft four times as powerful as that of men or quadrupeds, which is not
audible at more than half a league's diftance on
the furface. But this eftimation is even too
low: for, befide the diffipation of force to be
attributed to the caufe already affigned, the
                                    found

found is propagated in the higher regions as from a centre in all directions, and only a part of it reaches the ground; but, when made at the surface, the aerial waves are reflected as they roll along, and the lateral and vertical effect is augmented. It is hence that a person on the top of a tower hears one better at the bottom, than the person below hears from above.

Sweetness of voice and melody of song are qualities which in birds are partly natural, partly acquired. Their great facility in catching and repeating sounds enables them not only to borrow from each other, but often to copy the inflexions and tones of the human voice, and of our musical instruments. Is it not singular, that in all populous and civilized countries, most of the birds chant delightful airs, while, in the extensive deserts of Africa and America, inhabited by roving savages, the winged tribes utter only harsh and discordant cries, and but a few species have any claim to melody? Must this difference be imputed to the difference of climate alone? The extremes of cold and heat operate indeed great changes on the nature of animals, and often form externally permanent characters and vivid colours. The quadrupeds of which the garb is variegated, spotted, or striped, such as the panthers, the leopards, the zebras, and the civets, are all natives of the hottest climates. All the birds of the tropical regions sparkle with the most glowing tints,

while

while thofe of the temperate countries are ftain-
ed with lighter and fofter fhades. Of the three
hundred fpecies that may be reckoned belonging to our climates, the peacock, the common
cock, the golden oriole, the king-fifher, and the
goldfinch, only can be celebrated for the variety of their colours; but Nature would feem
to have exhaufted all the rich hues of the univerfe on the plumage of the birds of America,
of Africa, and of India. Thefe quadrupeds,
clothed in the moft fplendid robes, thefe birds
attired in the richeft plumage, utter at the fame
time hoarfe, grating, or even terrible cries.
Climate has no doubt a principal fhare in this
phænomenon; but does not the influence of
man contribute alfo to the effect? In all the do-
mefticated animals, the colours never heighten,
but grow fofter and fainter: many examples occur among the quadrupeds; and cocks and
pigeons are ftill more variegated than dogs or
horfes.   The real alteration which the human
powers have produced on nature, exceeds our
fondeft imagination: the whole face of the
globe is changed; the milder animals are tamed
and fubdued, and the more ferocious are repreffed and extirpated. They imitate our manners; they adopt our fentiments; and, under
our tuition, their faculties expand.   In the ftate
of nature, the dog has the fame qualities and
difpofitions, though in an inferior degree, with
the tiger, the leopard, or the lion; for the
                                    character

character of the carnivorous tribe results solely from the acuteness of their smell and taste : but education has mollified his original ferocity, improved his sagacity, and rendered him the companion and associate of man.

Our influence is smaller on the birds than on the quadrupeds, because their nature is more different from our own, and because they are less submissive and less susceptible of attachment. Those we call *domestic*, are only prisoners, which, but for propagating, are useless during their lives; they are victims, multiplied without trouble, and sacrificed without regret. As their instincts are totally unrelated to our own, we find it impossible to instil our sentiments; and their education is merely mechanical. A bird, whose ear is delicate, and whose voice is flexible, listens to discourse, and soon learns to repeat the words, but without feeling their force. Some have indeed been taught to hunt and fetch game; some have been trained to fondle their instructor : but these sentiments are infinitely below what we communicate so readily to the quadrupeds. What comparison between the attachment of a dog, and the familiarity of a canary bird; between the understanding of an elephant, and the sagacity of an ostrich ?

The natural tones of birds, setting aside those derived from education, express the various modifications of passion ; they change even according to the different times or circumstances.

ftances. The females are much more filent than the males ; they have cries of pain or fear, murmurs of inquietude or folicitude, efpecially for their young; but fong is generally withheld from them. In the male it fprings from fweet emotion, from tender defire ; the canary in his cage, the greenfinch in the fields, the oriole in the woods, chant their loves with a fonorous voice, and their mates reply in feeble notes of confent. The nightingale, when he firft arrives in the fpring, is filent ; he begins in faultering unfrequent airs : it is not until the dam fits on her eggs, that he pours out the warm melody of his heart : then he relieves and foothes her tedious incubation ; then he redoubles his careffes, and warbles more pathetically his amorous tale. And what proves that love is among birds the real fource of their mufic is, that, after the breeding feafon is over, it either ceafes entirely, or lofes its fweetnefs.

This melody, which is each year renewed, and which lafts only two or three months during the feafon of love, and changes into harfh low notes on the fubfidence of that paffion, indicates a phyfical relation between the organs of generation and thofe of voice, which is moft confpicuous in birds. It is well known that the articulation is never confirmed in the human fpecies before the age of puberty ; and that the bellowing of quadrupeds becomes tremendous when they are actuated by their fiery lufts. The repletion

pletion of the fpermatic veffels irritates the parts of generation, and by fympathy affects the throat. Hence the growth of the beard, the forming of the voice, and the extenfion of the genital organ in the male; the fwell of the breafts, and the expanfion of the glandulous bodies in the female. In birds the changes are more confiderable; not only are thefe parts ftimulated or altered; after being in appearance entirely deftroyed, they are even renovated by the operation of the fame caufes. The tefticles, which in man and moft of the quadrupeds remain nearly the fame at all times, contract and wafte almoft entirely away in birds after the breeding feafon is over, and on its return they expand to a fize that even appears difproportioned. It would be curious to difcover if there is not fome new production in the organs of the voice, correfponding to this fwell in the parts of generation.

Man feems even to have given a direction to love, that appetite which Nature has the moft deeply implanted in the animal frame. The domeftic quadrupeds and birds are almoft conftantly in feafon, while thofe which roam in perfect freedom are only at certain ftated times ftimulated by the ardour of paffion. The cock, the pigeon, and the duck, have, equally with the horfe, the ram, and the dog, undergone this important change of conftitution.

But the birds excel the other animals in the powers of generation, and in their aptitude for

motion. Many species scarcely rest a single moment, and the rapacious tribes pursue their prey
without halting or turning aside, while the quadrupeds need to be frequently recruited.—To
give some idea of the rapidity and continuance
of the flight of birds, let us compare it with the
celerity of the fleetest land-animals. The stag,
the rein-deer, and the elk, can travel forty
leagues a-day; the rein-deer can draw its sledge
at the rate of thirty leagues for several days.
The camel can perform a journey of three hundred leagues in eight days. The choicest race-
horse can run a league in six or seven minutes;
but he soon slackens his career, and could not
long support such an exertion. I have elsewhere
mentioned the instance of an Englishman who
rode sixty-two leagues in eleven hours and
thirty-two minutes, changing horses twenty-one
times: so that the best horse could not travel more
than four leagues in an hour, or thirty leagues
in a day. But the motion of birds is vastly
swifter: an eagle, whose diameter exceeds four
feet, rises out of sight in less than three minutes,
and therefore must fly more than 3,500 yards
in one minute, or twenty leagues in an hour.
At this rate, a bird would easily perform a journey of two hundred leagues in a day, since ten
hours would be sufficient, which would allow
frequent halts, and the whole night for repose.
Our swallows, and other migratory birds, might
therefore reach the equator in seven or eight

days. Adanfon faw on the coaft of Senegal
fwallows that had arrived on the ninth of October; that is, eight or nine days after their departure from Europe *. Pietro della Valle fays,
that in Perfia † the meffenger-pigeon travels as
far in a fingle day as a man can go a-foot in fix
days. It is a well-known ftory, that a falcon of
Henry II. which flew after a little buftard at
Fontainbleau, was caught next morning at Malta, and recognized by the ring which it wore ‡.
A Canary falcon, fent to the duke of Lerma,
returned in fixteen hours from Andalufia to the
ifland of Teneriffe, a diftance of two hundred
and fifty leagues. Sir Hans Sloane ‖ affures us,
that at Barbadoes the gulls make excurfions in
flocks to the diftance of more than two hundred
miles, and return the fame day. Taking all
thefe facts together, I think we may conclude
that a bird of vigorous wing could every day
pafs through four or five times more fpace than
the fleeteft quadruped.

Every thing confpires to the rapidity of a
bird's motion : firft, the feathers are very light,
have a broad furface, and their fhafts are hollow : fecondly, the wings are convex above and
concave below ; they are firm and wide fpread,
and the mufcles which act upon them are power-

* Voyage au Senegal.
† Voyage de Pietro della Valle.
‡ Obfervations of Sir Edmund Scoty, in Purchafs's Collection.
‖ A Voyage to the Weft Iflands, with their Natural Hiftory,
by Sir Hans Sloane,

ful :

ful: thirdly, the body is proportionally light, for the flat bones are thinner than in the quadrupeds, and hollow bones have much larger cavities. " The fkeleton of the pelican," fay the anatomifts of the Academy, " is extremely light, not weighing more than twenty-three ounces, though it is of confiderable bulk." This quality diminifhes the fpecific gravity of birds.

Another confequence which feems to refult from the texture of the bones, is the longevity of birds. In man and the quadrupeds, the period of life feems to be in general regulated by the time required to attain the full growth: but in birds it follows different proportions; their progrefs is rapid to maturity; fome run as foon as they quit the fhell, and fly fhortly afterwards: a cock can copulate when only four months old, and yet does not acquire his full fize in lefs than a year. Land animals generally live fix or feven times as long as they take to reach the age of puberty; but in birds the proportion is ten times greater, for I have feen linnets fourteen or fifteen years old, cocks twenty, and parrots above thirty, and they would probably go beyond thefe limits *. This difference

---

* A perfon of veracity affured me, that a parrot layed at about forty years of age, without commerce with any male, at leaft of its own kind.—It is faid, that a fwan has lived three hundred years; a goofe eighty; and a pelican as many. The eagle and crow are famous for longevity. ENCYCLOPEDIE, article *Oifean.*—Aldrovandus relates, that a pigeon lived twenty-two years, and ceafed to breed only the laft fix years.—Willoughby fays, that linnets live fourteen years, and goldfinches twenty-three, &c.

I fhould

I ſhould attribute to the ſoft porous quality of the bones; for the general oſſification and rigidity of the ſyſtem to which animals perpetually tend, determine the boundary of life; that will therefore be prolonged, if the parts want ſolidity and conſiſtence. It is thus that women arrive oftener at old age than men; that birds live longer than quadrupeds, and that fiſhes live longer than birds.

But a more particular inquiry will evince that uniformity of plan which prevails through nature. The birds, as well as the quadrupeds, are carnivorous, or granivorous. In the former claſs, the ſtomach and inteſtines are proportionally ſmall; but thoſe of the latter have a craw additional, correſponding to the falſe belly in ruminating animals, and the capacity of the ventricle compenſates for the unſubſtantial quality of their deſtined food. The granivorous birds have alſo two *cæca*, and a very ſtrong muſcular ſtomach, which ſerves to triturate the hard ſubſtances which they ſwallow.

The diſpoſitions and habits of animals depend greatly on their original appetites. We may therefore compare the eagle, noble and generous, to the lion; the vulture, cruel and inſatiable, to the tiger; the kite, the buzzard, the crow, which only prowl among carrion and garbage, to the hyænas, the wolves, and jackals. The falcons, the ſparrow-hawks, the gos-hawks, and the other birds trained for ſport, are analo-

gous

gous to the dogs, the foxes, the ounces, and the lynxes; the owls, which prey in the night, reprefent the cats; the herons, and the cormorants, which live upon fifh, correfpond to the beavers and otters; and, in their mode of fubfiftence, the woodpeckers refemble the anteaters. The common cock, the peacock, the turkey, and all the birds furnifhed with a craw, bear a relation to the ox, the fheep, the goat, and other ruminating animals. With regard to the article of food, birds have a more ample latitude than quadrupeds; flefh, fifh, the amphibious tribes, reptiles, infects, fruits, grain, feeds, roots, herbs; in a word, whatever lives or vegetates. Nor are they very nice in their choice, but often catch indifferently at what they can moft eafily obtain. The fenfe of tafte is much lefs acute in birds than in quadrupeds; for, if we except fuch as are carnivorous, their tongue and palate are in general hard, and almoft cartilaginous. Smell can alone direct them, and this they poffefs in an inferior degree. The greater number fwallow without tafting, and maftication, which conftitutes the chief pleafure in eating, is entirely wanting to them. Hence, on all thefe accounts, they are fo little attentive to the felection of their food, that they often poifon themfelves *.

* Parfley, coffee, bitter almonds, &c. prove poifonous to hens, parrots, and many other birds, which eat thefe fubftances with avidity when prefented with other food.

The

The attempt is impoffible therefore to diftinguifh the winged tribes according to the nature
of their aliments. The more conftant and determined appetites of quadrupeds might countenance fuch a divifion * ; but in birds, where the
tafte is fo irregular, it would be entirely nugatory. We fee hens, turkies, and other fowls
which are called granivorous, eat worms, infects, and bits of flefh with greater avidity than
grain. The nightingale, which lives on infects,
may be fed with minced meat ; the owls, which
are naturally carnivorous, often when other prey
fails, catch night-flies in the dark ; nor is their
hooked bill, as thofe who deal in final caufes
maintain, any certain proof that they have a
decided propenfity for flefh, fince parrots and
many other birds which feem to prefer grain

---

* Frifch, whofe work is in many refpects valuable, divides all
birds into twelve claffes. The firft contains *the fmall birds, with a
thick fhort bill, which fplit feeds into two equal portions* ; the fecond
includes *the fmall birds with a flender bill, that eat flies and worms* ;
the third comprehends *the black-birds and thrufhes* ; the fourth, *the
woodpeckers, cuckoos, hoopoes, and parrots* ; the fifth, *the jays and
magpies* ; the fixth, *the rooks and crows* ; the feventh, *the diurnal
birds of prey* ; the eighth, *the nocturnal birds of prey* ; the ninth, *the
wild and tame poultry* ; the tenth, *the wild and tame pigeons* ; the
eleventh, *the geefe, ducks, and other fwimming animals* ; the twelfth,
the birds which are fond of water and wet places.——We eafily fee
that the inftinct of opening feeds in two equal portions ought not
to be adopted as a character, fince in this fame clafs there are birds,
fuch as the titmice, that do not fplit them, but pierce and tear them ;
and that, befides, all the birds of this firft clafs, which are fuppofed
to fubfift folely on feeds, feed likewife on infects and worms : it was
better, therefore, as Linnæus has done, to join them into one clafs.

have

have alfo a hooked bill. The more voracious kinds devour fifh, toads, and reptiles, when they cannot obtain flefh. Almoft all the birds which appear to feed upon grain, were reared by their parents with infects. The arrangement derived from the nature of the food is thus totally deftitute of foundation. No one character is fufficient : it requires the combination of many.

Since birds cannot chew, and the mandibles which reprefent the jaws are unprovided with teeth, the grains are fwallowed whole, or only half-bruifed *. But the powerful action of the ftomach ferves them inftead of maftication ; and the fmall pebbles, which affift in trituration, may be conceived to perform the office of teeth †.

As

---

* In parrots, and many other birds, the upper mandible is moveable as well as the under ; whereas in quadrupeds the lower jaw only is moveable.

† In no animals is the mode of digeftion fo favourable as in birds to the fyftem of trituration. Their gizzard has the proper force and direction of fibres ; and the voracious kinds, which greedily fnatch the feeds on which they feed without ftopping to feparate the hard cruft which envelopes them, fwallow at the fame time little ftones, by means of which the violent contraction of the coats of the ftomach bruifes and detaches the fhell. This is a real trituration, which in other animals is performed by the teeth. But, after the feeds are decorticated, the action of a folvent may take place ; and there is a fort of bag from which a large quantity of a whitifh liquor flows into the ftomach, for in a recently dead bird it may be preffed out. Helvetius fubjoins, that fometimes in

the

As Nature has invefted the quadrupeds which haunt marfhes, or inhabit cold countries, with a double fur, and with thick clofe hair; fo has fhe clothed the aquatic birds, and thofe which live in the northern tracts, with abundance of plumage, and a fine down; infomuch that, from this circumftance alone, we may judge of their proper element, or of their natal region. In all climates, the birds which dwell in the water are nearly equally feathered, and have under the tail large glands, containing an oily fubftance for anointing their plumes, which, together with their thicknefs, prevents the moifture from infinuating. Thefe glands are much fmaller in the land-birds, or totally wanting.

Birds that are almoft naked, fuch as the oftrich, the caffowary, and the dodo, occur only in the warm climates. All thofe which inhabit cold countries are well clothed with plumage. And for the fame reafon, thofe which foar into the higher regions of the atmofphere require a thick covering, that they may encounter the

the *œfophagus* of the cormorant, fifh are found half digefted. *Hift. de l' Academie des Sciences, année* 1719.

Seventy *doubles* were found in the ftomach of an oftrich, moft of them worn three-fourths, and furrowed by their rubbing againft each other, and againft the pebbles, but not at all affected by folution, for fome which happened to be crooked were quite polifhed on the convex fide, while the concave fide was not altered. *Memoires pour fervir a l' Hiftoire des Animaux.*

A Spanifh gold piftole fwallowed by a duck had loft fixteen grains of its weight when voided. *Collect. Acad. Partie Etrangere.*

chilnefs

chilnefs which there prevails. If we pluck the feathers from the breaft of an eagle, he will no longer rife out of our fight.

The greater number of birds caft their feathers every year, and appear to fuffer much more from it than the quadrupeds do from a fimilar change. The beft fed hen ceafes at that time to lay. The organic molecules feem then to be entirely fpent on the growth of the new feathers. The feafon of moulting is generally the end of fummer or autumn *, and their feathers are not completely reftored till the beginning of fpring, when the mildnefs of the air, and the fuperabundance of nutrition, urge them to love. Then all the plants fhoot up, the infects awaken from their long flumber, and the earth fwarms with animation. This ample provifion fofters their ardent paffions, and offers abundant fubfiftence to the fruits of their embrace.

We might deem it as effential to the bird to fly, as it is to the fifh to fwim, or to the quadruped to walk; yet in all thefe tribes there are exceptions to the general property. Among

---

* Domeftic fowls generally moult in autumn; partridges and pheafants, before the end of fummer; and fuch as are kept in parks, caft their feathers immediately after their firft hatch. In the country, the pheafants and partridges undergo that change about the clofe of July, only the females which have had young are fome days later. Wild ducks moult rather before that time.——I owe thefe remarks to M. Le Roy, king's ranger at Verfailles.

quadru-

quadrupeds the rufous, red and common bats,
can only fly ; the feals, the fea-horfes, and fea-
cows, can only fwim ; and the beavers and
otters walk with more difficulty than fwim :
and, laftly, there are others, fuch as the floth,
which can hardly drag along their bodies. In
the fame manner, we find among birds the oftrich, the caffowary, the dodo, the touyou, &c.
which are incapable of flying, and are obliged
to walk ; others, fuch as the penguins, the fea-
parrots, &c. which fly and fwim, but never
walk ; and others, in fine, which, like the bird
of paradife, can neither walk nor fwim, but are
perpetually on the wing. It appears, however,
that water is, on the whole, more fuited to the
nature of birds than to that of quadrupeds : for,
if we except a few fpecies, all the land animals
fhun that element, and never fwim, unlefs they
are urged by their fears or wants. Of the birds,
on the contrary, a large tribe conftantly dwell
on the waters, and never go on fhore, but for
particular purpofes, fuch as to depofite their
eggs, &c. And what proves this pofition, there
are only three or four quadrupeds which have
their toes connected by webs ; whereas we may
reckon above three hundred birds which are
furnifhed with fuch membranes. The lightnefs
of their feathers and of their bones, and even
the fhape of their body, contribute greatly to
the facility with which they fwim, and their
feet ferve as oars to impel them along. Accordingly,

ingly, certain birds difcover an early propenfity
to the water; the ducklings fail on the furface
of the pool long before they can ufe their
wings.

In quadrupeds, efpecially thofe which have
their feet terminated by hard hoofs or nails, the
palate feems to be the principal feat of touch as
well as of tafte.   Birds, on the other hand,
oftener feel bodies with their toes; but the infide of thefe is covered with a callous fkin, and
their tongue and mouth are almoft cartilaginous: fo that, on both accounts, their fenfations
muft be blunt.

Such then is the order of the fenfes which
Nature has eftablifhed in the different beings.
In man, touch is the firft, or the moft perfect;
tafte the fecond; fight the third; hearing the
fourth; and fmell the fifth and laft.   In quadrupeds, fmell is the firft; tafte the fecond, or rather
thefe two fenfes form only one; fight the third;
hearing the fourth; and touch the laft. In birds,
fight is the firft; hearing the fecond; touch the
third; and tafte and fmell the laft.   The predominating fenfations will alfo follow the fame
order: man will be moft affected by touch;
the quadrupeds by fmell; and the birds by
fight.   Thefe will likewife give a caft to the
general character, fince certain motives of action will acquire peculiar force, and gain the
afcendency.   Thus, man will be more thought-
ful and profound, as the fenfe of touch would
appear

appear to be more calm and intimate ; the quadrupeds will have more vehement appetites; and the birds will have emotions as extenfive and volatile as is the glance of fight.

But there is a fixth fenfe, which, though it intermits, feems, while it acts, to control all the others, and excites the moft powerful emotions, and awakens the moft ardent affections:— it is love. In quadrupeds, that appetite produces violent effects ; they burn with maddening defire ; they feek the female with favage ardor ; and they embrace with furious extafy. In birds it is a fofter, more tender, and more endearing paffion ; and, if we except thofe which are degraded by domeftication, and a few other fpecies, conjugal fidelity and parental affection are among them alike confpicuous. The pair unite their labours in preparing for the accommodation of their expected progeny ; and, during the time of incubation, their participation of the fame cares and folicitudes continually augments their mutual attachment. After the eggs are hatched, a new fource of pleafure opens to them, which further ftrengthens the ties of affection ; and the tender charge of rearing the infant brood requires the joint attention of both parents. The warmth of love is thus fucceeded by calm and fteady attachment, which by degrees extends, without fuffering any diminution, to the rifing branches of the family.

The

The quadrupeds are impelled by unbridled luft, which never foftens into generous friend-fhip. The male abandons the female as foon as the cravings of his appetite are cloyed; he re-tires to recruit his ftrength, or haftens to the embraces of another. The education of the young is devolved entirely on the female; and as they grow flowly, and require her immediate pro-tection, the maternal tendernefs is ripened into a ftrong and durable attachment. In many fpe-cies the mother leads two or three litters at one time. There are fome quadrupeds, however, in which the male and female affociate together; fuch are the wolves and foxes: and the fallow-deer have been regarded as the patterns of conjugal fidelity. There are alfo fome fpecies of birds where the cock feparates after fatis-fying his paffion;—but fuch inftances are rare, and do not affect the general law of na-ture.

That the pairing of birds is founded on the need of their mutual labours to the fupport of the young, appears clearly from the cafe of the domeftic fowls. The male ranges at will among a feraglio of fubmiffive concubines; the feafon of love has hardly any bounds; the hatches are frequent and tedious; the eggs are often re-moved; and the female never feeks to breed, until her prolific powers are deadened, and al-moft exhaufted: befides, they beftow little care

in

in making their neft, they are abundantly fupplied with provifions, and by the affiftance of
man they are freed from all thofe toils and hard-
fhips and folicitudes which other birds feel and
fhare in common. They contract the vices
of luxury and opulence, *indolence* and *de-
bauchery*.

The eafy comfortable condition of the domeftic fowls, and their generous food, mightily
invigorate the powers of generation. A cock
can tread twelve or fifteen hens, and each embrace continues its influence for three weeks ;
fo that he may each day be the father of three
hundred chickens. A good hen lays a hundred
eggs between the fpring and autumn ; but in
the favage ftate fhe has only eighteen or twenty,
and that only during a fingle feafon. The
other birds indeed repeat oftener their incubations, but they lay fewer eggs. The pigeons,
the turtles, &c. have only two ; the great birds
of prey three or four ; and moft other birds five
or fix.

Want, anxiety, and hard labour, check in all
animals the multiplication of the fpecies. This
is particularly the cafe with birds ; they breed
in proportion as they are well fed, and afforded
eafe and comfort. In the ftate of nature, they
feem even to hufband their prolific powers, and
to limit the number of their progeny to the pe-
nury of their circumftances. A bird lays five
eggs, perhaps, and devotes her whole attention
during

during the reft of the feafon to the incubation
and education of the young. But if the neft
be deftroyed, fhe foon builds another, and lays
three or four eggs more; and if this be again
plundered, fhe will conftruct a third, and lay
ftill two or three eggs. During the firft hatch,
therefore, thofe internal emotions of love which
occafion the growth and exclufion of the eggs,
are repreffed. She thus facrifices duty to paffion, amorous defire to parental attachment.
But when her fond hopes are difappointed, fhe
foon ceafes to grieve; the procreative faculties,
which were fufpended, not extinguifhed, again
refume their influence, and enable her in fome
meafure to repair her lofs.

As love is a purer paffion in birds than in
quadrupeds, its mode of gratification is alfo
fimpler. Coition is performed among them
only in one way*, while many other animals
embrace in various poftures †: only in fome
fpecies, as in that of the common cock, the female fquats; and in others, fuch as the fparrows, fhe continues to ftand erect. In all of
them the act is tranfitory, and is ftill fhorter in
thofe which in their ordinary attitude wait the
approach of the male, than in thofe which
cower to receive him ‡. The external form,

* Ariftotle, lib. v. 8.

† The fhe-camel fquats; the fhe-elephant turns upon her back;
the hedgehogs couple face to face, and either in an erect or re-
clined pofture; and monkies in every manner.

‡ Ariftotle, lib. v. 2.

and

and the internal ſtructure of the organs of generation are very different from what obtains
in quadrupeds. The ſize, the poſition, the number, the action and motion of theſe parts even
vary much in the ſeveral ſpecies of birds *. In
ſome there appears to be a real penetration ; in
others, a vigorous compreſſion, or ſlight touch.
But we ſhall conſider the details in the courſe
of the work.

To concentrate the different principles eſtabliſhed in this diſcourſe : that the *ſenſorium* of
birds contains chiefly the images derived from the
ſenſe of ſight ; and theſe, though ſuperficial, are
very extenſive, and, for the moſt part, relate to
motion, to diſtance, and to ſpace : that comprehending a whole province within the limits
of their horizon, they may be ſaid to carry in
their brain a geographical chart of the places
which they view : that their facility in tra-
verſing wide territories is one of the cauſes
which prompt their frequent excurſions and
migrations : that their ear being delicate, they
are alarmed by ſudden noiſes, but may be ſooth-
ed by ſoft ſounds, and allured by calls : that
their organs of voice being exceedingly power-

* Moſt birds have two yards, or a forked one projecting from
the *anus*. In ſome ſpecies the male organ is exceedingly large ;
in others hardly viſible. The female orifice is not ſituated, as in
the quadrupeds, below the *anus*, but above it ; and there is no ma-
trix, &c.

ful

ful and foft, they naturally vent their feelings in loud refounding ftrains : that, as they have more figns and inflexions, they can, better than the quadrupeds, exprefs their meaning : that eafily receiving, and long retaining the imprefsions of founds, the organ delights in repeating them ; but that its imitations are entirely mechanical, and have no relation to their conceptions : that their fenfe of touch being obtufe, they have only imperfect ideas of bodies : that they receive their information of diftant objects from fight, not from fmell : that as their tafte is indifcriminating, they are more prone to voracity than fenfuality : that, from the nature of the element which they inhabit, they are independent of man, and retain their natural habits ; that, for this reafon, moft of them are attached to the fociety of their fellows, and eagerly convene : that, being obliged to unite their exertions in building a neft, and in providing for their offspring, the pair contract an affection for each other, which continues to grow, and then extends to the tender brood : that this friendfhip reftrains the violent pafsions, and even tempers love, and begets chaftity, and purity of manners, and gentlenefs of difpofition : that, though their power of fruition is greater than in other animals, they confine its exercife within moderate bounds, and

ever

ever fubject their pleafures to their duties:
and, finally, that thefe fprightly beings, which
Nature would feem to have produced in her
gay moments, may be regarded as a ferious
and decent race, which exhibit excellent lef-
fons and laudable examples of morality.

EXPLANATION *of some* TECHNICAL TERMS
that occur in this Work.

Mandible, one of the pieces of which the bill confifts.
Vent, the part under the tail.
Cere, the naked fkin which covers the bafe of the bill in fome
    birds; fo called from its refembling wax.
Bridle, the plumules on the front immediately over the bill.
Strap, the fpace running from the bill to the eye.
Orbit, the naked fkin encircling the eye.
Quill, a great feather of the wings or tail.
Rufous, tawny-red.
Fulvous, tawny yellow.
Cinereous, afh-coloured, rather deep.
Ferruginous, dark, rufty-coloured.

———————

The Meafures and Weights ufed throughout are French. The
Parifian foot is to the Englifh as 1 is to 1.066: hence the following table is conftructed.

Inches.		Inches.		Inches.	
French.	Englifh.	French.	Englifh.	French.	Englifh.
4 -	4.26	13 -	13.85	22 -	23.46
5 -	5.33	14 -	14.92	23 -	24.52
6 -	6.40	15 -	15.99	24 -	25.58
7 -	7.46	16 -	17.05	25 -	26.65
8 -	8.53	17 -	18.12	26 -	27.72
9 -	9.59	18 -	19.18	27 -	28.78
10 -	10.66	19 -	20.25	28 -	29.85
11 -	11.73	20 -	21.32	29 -	30.91
12 -	12.79	21 -	22.38	30 -	31.98

The Parifian pound is divided into fixteen ounces, each ounce
into eight gros, and each gros into feventy-two grains. The
pound is equal to 7561 Englifh grains Troy; whence the French
ounce amounts to 472½ grains Troy; the gros to 59 grains, and
a French grain is about four-fifths of an Englifh grain. A French
ounce is therefore only one fixty-fourth greater than an ounce
Troy, which makes it unneceffary to give a table of reduction.

BIRDS of PREY.

A L L the birds almoſt might merit this appellation, ſince by far the greater number
ſearch for infects, worms, and other ſmall creatures; but I ſhall confine it to thoſe which ſub-
ſiſt on fleſh, and wage perpetual war againſt the
other winged tribes. On compariſon, I find that
they are much leſs numerous than the ravenous
quadrupeds. The family of the lions, the tigers,
the panthers, the ounces, the leopards, the hunting cats, the jaguars, the couguars, the Mexican
cats, the margays, and the wild or domeſtic cats :
that of the dogs, the jackals, the wolves, the foxes,
and the arctic foxes : the more numerous tribes
of the hyænas, the civets, the oriental civets, the
dwarf civets, the Madagaſcar pole-cats : the ſtill
more numerous tribes of the pole-cats, the mar-
tens, the fitchews, the ſkunks, the ferrets, Guinea
weaſels, the ermines, the common weaſels, the
ſables, the ichneumons, the Braſilian weaſels, the
gluttons, the pekans, the minks, the ſouſliks ;
the opoſſums, the mar-mice, the Mexican opoſſums, the woolly jerboas, the Surinam opoſſums :
that of the rufous, red, and common bats : To
theſe we may add the whole family of the rats,

which

which being too weak to attack other animals, prey on each other:—all thefe rapacious quadupeds exceed greatly in number the eagles, the vultures, the fparrow hawks, the falcons, the jerfalcons, the kites, the buzzards, the keftrels, the merlins, the owls, the fhrikes, and the crows, which are prone to rapine : and many of thefe, fuch as the kites, the buzzards, and the crows, prefer carrion to frefh prey. In fhort, there is only a fifteenth part of the birds carnivorous, while of the quadrupeds more than a third come under that defignation.

The birds of prey being much fewer and weaker than the rapacious quadrupeds, commit lefs depredation on land ; but, as if tyranny never relinquifhed its claims, whole tribes inhabit the ocean and fubfift by their ravages. Of the quadrupeds, fcarce any, except the beavers, the otters, the feals, and the fea-horfes, live on fifh ; yet multitudes of birds derive their fupport entirely from that fource. We have therefore to divide the birds of prey into two claffes, correfponding to the elements of air and water, which are the fcenes of their havocks. Thofe which war againft the finny race are provided with a ftraight pointed bill; their nails are flender, their toes webbed, and their legs bent backwards. Thofe, on the contrary, which riot in carnage at land, and which are properly the fubject of this article, are furnifhed with talons and with a fhort curved bill; their toes are parted, and
without

without membranes; their legs are ſtrong, and generally covered by the feathers of the thighs; their nails large and hooked.

We ſhall for the preſent ſet aſide alſo the nocturnal birds of prey, and adopt what appears to be the moſt natural order in treating of thoſe which commit their ravages during the day. We ſhall begin with the eagles, the vultures, the kites, and the buzzards; then the hawks, the jerfalcons, and falcons; and cloſe with the merlins and the ſhrikes. Many of theſe include a great number of ſpecies and of permanent families produced by the influence of climate; and with each we ſhall range the kindred foreign birds. In this way we ſhall delineate not only thoſe of Europe, but alſo all thoſe which inhabit remote countries, whether deſcribed by authors, or procured by our correſpondence.

There is a ſingular property common to all the birds of prey, but of which it would be difficult to aſſign the cauſe *; that the female is ſtronger, and a third larger than the male; exactly the reverſe to what obtains in the quadrupeds, and even in other birds. In fiſhes and inſects, the female is indeed larger than the male: this is

---

* The final cauſe at leaſt is obvious. In the rapacious birds, the care of the brood is entruſted ſolely to the female; and Nature has wiſely endowed her with greater force to enable her to provide both for her own wants and thoſe of her family.—Nothing but the too frequent and often puerile applications of ſuch views of Nature could ever bring them into diſrepute. T.

owing

owing to the immenfe number of eggs which fwell their bodies. But this reafon will not apply in the cafe of birds.—In thofe which are the moft prolific, fuch as the domeftic poultry, the ducks, turkies, pheafants, partridges, and quails, the hen lays eighteen or twenty eggs, and yet is fmaller than the cock.

All the birds of rapine fly in a lofty courfe, their wings and legs are ftrong, their fight exceedingly quick, their head thick, their tongue flefhy, their ftomach fingle and membranous, their inteftines narrower and fhorter than in other birds; they prefer the folitary tracts, the defert mountains, and they commonly breed in crags, or on the talleft trees. Many fpecies inhabit both continents, and fome appear to have no fixed abode. The general characters are, that their bill is hooked, and that they have four toes on each foot, all of which are diftinctly parted. But the eagle's head is covered with feathers, which diftinguifhes it from the vulture, whofe head is naked, and only fhaded with flight down: And both thefe are difcriminated from the hawks, the buzzards, the kites, and the falcons, by an obvious property; for their bill continues ftraight to a certain diftance before it bends, but in the latter it affumes its curve at the origin.

The birds of prey are not fo prolific as other birds. It is ftrange that Linnæus fhould affert that they lay about four eggs: for there are fome, fuch as

the

the common and fea-eagles, which have only two;
and others, as the keftrel and merlin, that have
feven. In birds, as in quadrupeds, the general
law obtains, that the multiplication is inverfely
as the bulk. There are fome apparent exceptions to this rule, pigeons for inftance; but the
fmallnefs of the hatch will be found to be compenfated by its frequent repetition.

The birds of prey are more obdurate and ferocious than other birds. They are not only
intractable, but have the unnatural propenfity to
drive their tender brood from the neft. Accuf-
tomed continually to fcenes of carnage, and
torn by angry paffions, they contract a ftern
cruel difpofition; all the fofter feelings are era-
dicated, and maternal attachment itfelf is blunted.
She regards not the imploring calls of her helplefs young, but when ftraitened for food, fhe
rudely thrufts them upon the world, or murders
them in a tranfport of fury.

This obdurate felfifh temper produces in the
birds of prey, as well as the carnivorous quadrupeds, another effect. They never affociate
together, but, like robbers, lead a roving folitary
life. Luft indeed draws together the male and
female, and, as they can mutually affift in the
purfuit of prey, they feldom feparate even
after the breeding feafon. But the family never
coalefces; and the larger kinds, fuch as the eagle,
will not fuffer their young to be rivals, but expel them from their domain: Whereas, all birds

and

and quadrupeds which fubfift on the fruits of
the earth, live in harmony with their offspring,
or affemble joyoufly in numerous troops.

Before we proceed to the detail of facts, we
cannot avoid making fome remarks on the common methods of claffification. The nomencla-
tor ftrives to defcribe the colours of the plumage
with minute precifion ; he enumerates their difpofition, all the fhades, the fpots, the bars, the
ftripes, the lines ; and if a bird does not come
under the defcription which he has thus formed
he regards it as a different fpecies. But all animals change their early garb and complexion ;
and the tints of the rapacious birds are wonderfully altered by the firft moulting. A fecond
confiderable one fucceeds, and this is often followed by a third; fo that a perfon who fhould
judge entirely from the colours, would imagine
that a bird of fix months old, another of the
fame kind of eighteen months, and another of
two years and a half, belonged to three different
fpecies. But the plumage is alfo affected by
various other caufes ; by difference of fex, of
age, and of climate ; and therefore the colours
can never afford any permanent diftinction.

The EAGLES.

MANY birds come under this defignation. Our nomenclators reckon eleven fpecies natives of Europe, befides four other, two of which are from Brafil, one from Africa, and another from the Eaft Indies. Thefe eleven fpecies are : firft, the Common Eagle ; fecond, the White-headed Eagle ; third, the White Eagle; fourth, the Spotted Eagle ; fifth, the White-tailed Eagle ; fixth, the Little White-tailed Eagle ; feventh, the Golden Eagle ; eighth, the Black Eagle ; ninth, the Great Sea-eagle ; tenth, the Sea-eagle ; and, eleventh, the White John. Nothing is eafier than to fwell the catalogue of names, and by a profufion of divifions and diftinctions to dazzle the ignorant. We need only to wade through books, ranfack cabinets, and adopt as fpecific characters all the differences in fize or colour that may occur. But the true object of the naturalift is to weigh and reflect ; to endeavour to feize the general views, and to concentrate and arrange ; and thus, by introducing order and precifion, to fmooth the progrefs of the ftudent.

Omitting

Omitting therefore the four foreign species of eagles, which we shall consider in the sequel, and excluding from the list the *White John*, which is entirely a different bird, we may reduce the eleven species to six, of which there are three only that properly deserve the name of Eagles. These three are: first, the Golden Eagle; second, the Common Eagle; third, the Rough-footed Eagle. The remaining three are: first, the *Pygargue*, or Bald Eagle; second, the Osprey; third, the Sea-eagle.

The Golden and Rough-footed Eagles form each an independent and unconnected species; but the Common and Bald Eagles are subject to variety. The species of the Common Eagle includes the brown and the black. The Rough-footed Eagle contains three varieties, viz. the Great White-tailed Eagle, the Little White tailed Eagle, and the White-headed Eagle. I shall not add the White Eagle, for I am confident that it owes its colour to the influence of excessive cold.

I am induced to adopt this arrangement, both because it was known even in the time of the ancients, that the different kinds of Eagles intermix, and because it nearly coincides with the division marked by Aristotle, who appears to have been better acquainted than any of our nomenclators with the real discriminating characters. He says, that there are six species of Eagles; but among these he includes a bird, which he himself confesses belongs rather to

the

the vultures *, and which we muſt therefore ſet
aſide. Of the five remaining ones, the three
firſt are the ſame with thoſe on which I have
fixed ; and the fourth and fifth correſpond to
the Bald Eagle and the Oſprey. I have ventur-
ed, notwithſtanding the authority of that great
philoſopher, to ſeparate theſe laſt from the Eagles properly ſo called ; in other reſpects, our
ideas exactly correſpond.—I ſhall conſider theſe
ſubjects fully in the following articles.

* The fourth kind of Eagles, is the *Percnopterus,* ſo called on account of the ſpots on its wings ; its head is whitiſh, and its body is
larger than the three firſt, but its wings ſhorter, and its tail longer.
It has the aſpect of a vulture, which has procured it the epithets of
Half-eagle and Mountain Stork. This degenerate bird inhabits the
foreſts ; it has all the bad qualities of the others, but inherits none
of their generous diſpoſitions, for it is beaten and driven to flight
by the crow ; it is lean, hungry, and gaunt ; perpetually complaining, noiſy, and clamorous.

GOLDEN EAGLE.

Le Grand Aiglé, Buff.
Falco Chryfaëtos, Linn.
In Spanifh, *Aquila coronada.*
In Polifh, *Orzelprzedni.*
In Perfian, *An fi muger.*
In Syriac, *Napan.*
In Chaldaic, *Nifra.*
In Arabic and Hebrew, *Nefer.*

THE firft fpecies is the Golden Eagle, which Belon named, after Athenæus, the *Royal Eagle,* or the *King of Birds.* This is indeed an eagle of a noble family and of an independent race. Hence Ariftotle denominates it αετος γνησιος *(the Eagle of Birth),* and our nomenclators have named it the *Golden Eagle :* It is the largeft of the genus. The female meafures, from the point of the bill to the extremity of the feet, more than three feet and an half; the wings, when expanded, extend above eight feet, and it weighs fixteen or eighteen pounds. The male is fmaller and does not weigh more than twelve. In both, the bill is very ftrong, and refembles bluifh horn; the claws are black and pointed, and the one placed behind, which is the largeft, is fometimes five inches long : the eyes are large, but funk in a

deep

THE GOLDEN EAGLE

deep cavity, and covered by the projection of
the fuperior part of the orbit: the iris is of a
fine bright yellow, and fparkles with dazzling
fire; the vitreous humour is of a topaz colour;
the cryftalline lens, which is dry and folid, has
the luftre and brilliancy of the diamond: the
œfophagus dilates into a large bag, which
is capable of containing a pint: the ftomach,
which is under this, is not near fo large, but is
equally pliant and membranous. The bird is
plump, efpecially in winter. The fat is white,
and the flefh, though hard and fibrous, has not
that wild flavour common to birds of prey.

This fpecies inhabits Greece, the mountains of
Bugey in France, thofe of Silefia in Germany,
the forefts of Dantzic, the fummits of the Car-
pathian mountains, the Pyrenees, and the mountains of Ireland. It is found alfo in Afia Minor,
and in Perfia; for the Perfians had, before the
Romans, affumed the eagle as the ftandard of
war; and it was this great eagle, this golden
eagle, *aquila fulva*, which was confecrated to
Jupitez. The teftimony of travellers afcertains
its exiftence in Arabia, in Mauritania, and in
many other provinces of Africa and Afia, as far
as Tartary; but it has not been difcovered in
Siberia, or in any other part of the north of
Afia. The fame remark may be extended to
Europe. For this noble bird, which is every
where rare, is more frequent in the warm regions than in the temperate countries, and it is

feldom

feldom obferved to penetrate farther north-
wards than the latitude of fifty-five degrees.
Nor is it found in North America, though the
common eagle is an inhabitant of that part of
the globe. The Golden Eagle feems to have
continued its ancient refidence; like the other
animals, which, being unable to fupport an intenfe cold, could not migrate into the new
world.

There are feveral points, both phyfical and
moral, in which the eagle refembles the lion.
Both are alike diftinguifhed by their ftrength;
and hence the eagle extends his dominion over
the birds, as the lion over the quadrupeds.
Magnanimity is equally confpicuous in both;
they defpife the fmall animals, and difregard their
infults. It is only after a feries of provocations,
after being teazed with the noify and harfh notes
of the raven or magpie, that the eagle is determined to punifh their temerity or their infolence
with death. Befides, both difdain the poffeffion
of that property which is not the fruit of their
own induftry; and they rejeʤ with contempt
the prey which is not procured by their own
exertions. Both are remarkable for their tem-
perance. The eagle feldom devours the whole
of his game, but, like the lion, leaves the fragments and offals to the other animals. Though
famifhed for want of prey, he difdains to feed
upon carrion. Like the lion alfo, he is folitary,
the inhabitant of a defert, over which he reigns
supreme,

fupreme, and excludes all the other birds from
his filent domain. It is more uncommon perhaps to fee two pairs of eagles in the fame tract
of the mountain, than two families of lions in
the fame part of the foreft. They feparate from
each other at fuch wide intervals, as to afford
ample range for fubfiftence, and efteem the
value and extent of their kingdom to confift in
the abundance of the prey with which it is re-
plenifhed. The eyes of the eagle have the glare
of thofe of the lion, and are nearly of the fame
colour; the claws of the fame fhape, the organs
of found are equally powerful, and the cry is
equally terrible. Deftined both of them for war
and plunder, they are equally fierce, equally
bold, and intractable. It is impoffible to tame
them, unlefs they be caught when in their infancy. It requires much patience and art to
train a young eagle for the chace; and, after he
has attained to age and ftrength, his caprices
and momentary impulfes of paffion are fufficient
to create fufpicions and fears in his mafter.
Authors inform us, that the eagle was anciently
ufed in the eaft for falconry, but this practice is
now laid afide. He is too heavy to be carried
on the hand without great fatigue, nor is he
ever brought to be fo tame or fo gentle, as to
remove all fufpicions of danger. His bill and
claws are crooked and formidable: his figure
correfponds to his inftinct. His body is robuft;
his legs and wings ftrong; his flefh hard; his

bones

bones firm; his feathers ftiff; his attitude bold
and erect; his movements quick; his flight
rapid. He rifes higher in the air than any of
the winged race, and hence he was termed by the
ancients the *Celeftial Bird*, and regarded, in their
auguries, as the meffenger of Jupiter. He can
diftinguifh objects at an immenfe diftance, but
his fmell is inferior to that of the vulture. By
means of his exquifite fight, he purfues his prey,
and, when he has feized it, he checks his flight,
and places it upon the ground, to examine its
weight, before he carries it off. Though his
wings be vigorous, yet his legs being ftiff, it is
with difficulty that he can rife, efpecially if he
is loaded. He bears away geefe and cranes with
eafe; he alfo carries off hares, young lambs and
kids. When he attacks fawns or calves, he
inftantly gluts himfelf with their blood and
flefh, and afterwards tranfports the mangled
carcafes to his *eyry* or *airy*, (fo his neft is called,)
which is quite flat, and not hollow like that of
other birds. He commonly places it between
two rocks, in a dry inacceffible place. The fame
neft, it is faid, ferves the eagle for the whole
courfe of his life. It is indeed a work laborious enough not to be repeated, and folid enough
to laft for a confiderable time. It is conftructed
nearly like a floor, with fmall fticks, five or fix
inches long, fupported at the extremities, and
croffed with pliant branches, covered with feveral
layers of rufhes and heath: the neft is feveral
feet

feet broad, and fo firm, as not only to receive
the eagle, the female, and the young, but to
bear the weight of a large quantity of provifions.
It is not covered above, but is fheltered by the
projection of the upper part of the rock. In
the middle of this ftructure, the female depofites
her eggs, which feldom exceed two or three,
and covers them, it is faid, for thirty days; but
fome of thefe are commonly addle, and it is
feldom that three young eagles are found in a
fingle neft. It is even pretended, that after they
have acquired fome ftrength, the mother deftroys
the weakeft or the moft voracious of her infant
brood. Exceffive fcarcity of provifions alone
can occafion this unnatural treatment. The
parents, not poffeffing a fufficiency for their own
fupport, endeavour to reduce the members of
their family; and when the young are able to fly,
and in fome degree to provide for themfelves,
they expel them from their natal abode, and
never fuffer them to return.

The plumage is not of fo deep a caft in
the young eagles as in thofe that are full grown.
At firft it is white, then a faint yellow, and
afterwards it becomes a bright copper colour.
Age, as well as gluttony, difeafe, and captivity,
contributes to render them white. It is faid they
live above a century, and that their death is not
occafioned fo much by extreme age, as by the
inability to take food, the bill growing fo much

E 2   curved

curved as to become ufelefs. However, it has been obferved, that eagles kept in confinement occafionally fharpen their bill, and that its increafe is, for feveral years, imperceptible. It has alfo been remarked, that they feed upon every kind of flefh, and even upon that of other eagles. When they cannot procure flefh, they greedily devour bread, ferpents, lizards, &c. If they be not fupplied with food, they bite cruelly the cats, dogs, and men that come within their reach. At intervals, they pour forth in an equable ftrain their fhrill, loud, and lamentable notes.—The eagle drinks feldom, and perhaps not at all when in perfect liberty, becaufe the blood of his victims are fufficient to quench his thirft. His excrements are always foft, and more watery than thofe of the other birds, even thofe which drink frequently.

To this great fpecies we muft refer the account in the paffage of Leo Africanus which we have already quoted, and what travellers in Africa and Afia relate, who agree in afferting that this bird not only carries off kids and young deer, but when taught, that it will even attack foxes and wolves *.

* The Emperor of Thibet has feveral tame eagles, which are fo keen and fierce, that they feize hares, bucks, does, and foxes; and there are fome fo extremely bold, that they rufh impetuoufly upon the wolf, and harafs him fo much that he can be more eafily caught. MARCO POLO.

[A] Linnæus refers the eagles to the genus of the falcon. The specific character of the Golden Eagle is, " that its cere is yellowish, its feet woolly and rusty-coloured, its body of a dusky variegated ferruginous colour, the tail black, with a waved cinereous base." He adds, that its feet are clothed with feathers as far as the nails; and that in fine weather it soars into the aërial regions, but when there is an impending storm, it hovers near the earth.

E 3

The RING-TAIL EAGLE,

L'Aigle Commun, Buff.
Falco Fulvus, Linn.
Aquila, Briff. and Klein.
Chryfaëtos, caudâ annulo albo cinĉta, Will. and Ray.
The Black Eagle, Penn.
In Spanifh, *Aquila Conocida.*
In German, *Adler, Arn, Aar.*

THIS fpecies of eagle is not fo pure or generous as the Golden Eagle. It is compofed of two varieties; the brown eagle, and
the black eagle. Ariftotle has not diftinguifhed
them by name; and it appears that he claffed
them under the denomination of Μελαιναετος;
that is, black or blackifh eagle. He properly
feparates this fpecies from the preceding, becaufe
it differs : 1. in fize; the Ring-tail Eagle, whether black or brown, being fmaller than the
Golden Eagle : 2. by the colours, which are conftant in the Golden Eagle, but vary in the Ring-
tail Eagle : 3. by its cry, the Golden Eagle utter-
ing often a doleful plaint, while the Ring-tail
Eagle, black or brown, feldom fcreams : 4. by
its natural difpofitions; the Ring-tail Eagle feeding

ing all its young in the neft, training them, and conducting them to prey after they are partly grown; while the Golden Eagle drives them out of its airy, and abandons them as foon as they are able to fly.

It appears eafy to prove that the Brown and Black Eagle, which I have claffed together, do not really conftitute two diftinct fpecies. We need only compare them together, even from the characters given by nomenclators with the view of diftinguifhing them. They are both nearly of the fame fize; they are of the fame brown colour, only fometimes of a deeper fhade; in both, the upper part of the head and neck is tinged with ferruginous, and the bafe of the large feathers marked with white; the legs and feet are alike clothed; in both, the iris is of a hazel colour, the cere of a bright yellow, the bill that of bluifh horn, the toes yellow, and the talons black: in fhort, the whole difference confifts in the fhades and diftribution of the colour of the feathers; which is by no means fufficient to conftitute two different fpecies, efpecially when the number of the points of refemblance fo evidently exceeds that of the difference. I have therefore without fcruple reduced thefe two fpecies to one. Ariftotle has done the fame thing without mentioning it; but it appears that his tranflator, Theodore Gaza, perceived it; for he does not render Μελαιναέτος by *Aquila nigra*, but by *Aquila nigricans, pulla fulvia*, which includes the

E 4   two

two varieties of this fpecies, both of which are blackifh, but the one of which is more tinged with yellow than the other. Ariftotle, whofe accuracy I often admire, gives names and epithets to the animals which he mentions. The epithet of this bird is λαγωφονος, or the *deftroyer of hares*. In fact, though the other eagles alfo prey upon hares, this fpecies is a more fatal enemy to thofe timid animals, which are the conftant object of their fearch, and the prey which they prefer. The Latins, after Pliny, termed this eagle *Valeria, quafi valens viribus*, becaufe of its ftrength, which appears greater than that of the other eagles in proportion to the fize.

The Ring-tail Eagle is more numerous and fpread than the Golden Eagle. The latter is found only in the warm and temperate countries of the ancient continent; the former prefers the cold tracts, and inhabitants of both continents. It occurs in France, Savoy, Switzerland, Germany, Poland, Scotland, and even in North America, at Hudfon's Bay*.

---

* Ellis tells us, that about Hudfon's Bay there are many other birds remarkable for their fhape and ftrength : fuch as the White-tailed Eagle, which is nearly of the fize of a turkey-cock ; its crown flattened, its neck fhort, its breaft large, its thighs ftrong, and its wings very long and broad in proportion to its body ; they are blackifh behind, but alfo of a lighter colour on the fides ; the breaft is marked with white, the wing feathers are black ; the tail when clofed is white above and below, except the tips of the feathers, which are black or brown ; the thighs are covered with blackifh brown feathers, through which in fome places the white down

down appears ; the legs are covered to the feet with a brown, or somewhat reddiſh plumage; each foot has four thick ſtrong toes, three before and one behind; they are covered with yellow ſcales, and furniſhed with nails that are exceedingly ſtrong and ſharp, and of a ſhining black.

[A] The ſpecific charaćter given by Linnæus of the Ring-tail Eagle *(Falco fulvus)* is, " That its cere is yellow; its feet woolly and dull ruſt-coloured; and its tail marked with a white ring." The brown ſort was ſtated in the 10th edition of the *Syſtema Naturæ* as a diſtinćt ſpecies, by the name of *Falco Canadenſis*, and deſcribed as " having a yellow cere, its feet woolly, its body duſky-coloured, its tail white, and tipt with brown." In the 12th and ſubſequent editions, however, it is conſidered as merely a variety. It builds its neſt in the lofty cliffs. The ſpaces between its eyes and its ears are naked. Its breaſt is ſprinkled with triangular ſpots.

We may remark, that both Linnæus and Pennant conceive, that Marco Polo, in his Deſcription of the Uſages of the Tartars, alludes to this ſpecies, and not to the Golden Eagle, as Buffon ſuppoſes.

The Black Eagle, termed by Friſch, *Schwartz-braune Adler* (Black-brown Eagle), which Buffon ranges with the Ring-tail Eagle, is reckoned a different ſpecies by Linnæus, under the name of *Falco Melanæetus.*—" Its cere is yellowiſh, its feet partly woolly, its body black-ferruginous, with yellow ſtreaks." It is two feet ten inches long. The half of the wing feathers next their origin is white with blackiſh ſpots, the remaining half blackiſh. The egg is a dirty white, mottled with ruſty-clouded ſpots.

The ROUGH-FOOTED EAGLE.

Le Petit Aigle. Buff.
Falco Nævius. Linn.
In German, *Stein Adler, Gauſe aar.*

THE third ſpecies is the Rough-footed Eagle,
which Ariſtotle deſcribes as a plaintive bird,
with a ſpotted plumage, and ſmaller and weaker
than the other eagles. It meaſures, from the
point of the bill to the extremity of the feet,
only two feet and a half; and its wings are
proportionally ſmaller, ſcarcely extending four
feet. It has been termed *Aquila planga* [B],
Aquila clanga, the *Plaintive Eagle,* the *Screaming
Eagle.* Theſe names are very applicable; for it
continually utters moans, or lamentable cries.
It was ſurnamed *Anataria,* becauſe it commonly
preys upon ducks; *Morphna,* becauſe its plumage, which is of a dirty-brown, is marked upon
the thighs and wings with ſeveral white ſpots,
and its neck is encircled with a large whitiſh

[B] Syſtematic writers have conſidered the *Crying* or *Spotted* Eagle
as a different ſpecies from the Rough-footed. It is the *Falco Ma-
culatus* of Linnæus. The character: " Its cere and its woolly feet
are yellowiſh, its body duſky-ferruginous below; the axillary feathers and the coverts of the wings are tipt with oval white ſpots."
It is two feet long.

ring.

ring. It is more tractable * than any of the
eagles, and not so bold or intrepid. It is termed by the Arabians *Zemiech* †, to distinguish it
from the Golden Eagle, which is called *Zumach*.
The crane is its largest prey, and it generally
confines its ravages ‡ to the ducks, the small
birds and rats. This species, though not plentiful in any particular spot, is scattered over
the extent of the ancient continent § ; but it
does not appear that it is found in America :
for I presume that the bird called the Oronooko
Eagle, which bears some resemblance to this in
the variety of its plumage, is yet of a different
species.—If this Rough-footed Eagle, which is
much more docile, and more easily tamed than
the other two, and which is also lighter on the
hand, and less dangerous to its master, were
equally intrepid, it would have been employed
for the purposes of falconry. But it is as cowardly as it is plaintive and noisy. A well-trained
sparrow-hawk can attack it, and come off vic-

* This *Aquila clanga* lived familiarly with me for more than
three years. It would, when I allowed it, sit upon the table several
hours at my left-hand, observing the motion of the right in writing, and sometimes stroked my cap with its head. If I tickled
it under the chin, it uttered a shrill sound. It lived peaceably with
the other birds. It disliked every food but fresh beef. KLEIN.

† The *Zumach* preys upon hares, foxes, and deer ; the *Zemiech*
catches cranes, and the smaller birds. *Falconnerie, par* GUIL.
TARDIF.

‡ Schwenckfeld.

§ It is found near Dantzic ; and also, though rarely, in the
mountains of Silesia. SCHWENCKFELD.

torious.

torious. Befides, our authors on the fubject of
falconry inform us, that, in France at leaft, the
two firft fpecies of eagles only have been trained
for fport*. To fucceed in teaching them, they
muft be taken when young, for an adult eagle
is not only ftubborn, but quite intractable. They
muft be fed upon the flefh of the game which
they are intended to purfue. Their education
requires more watchful attention than that of
the other birds employed in falconry.—We fhall
give a fketch of that art when we treat of the
falcon. I fhall only mention here fome peculiarities which have been obferved with regard to
eagles, whether in the ftate of liberty, or in that
of domeftication.

The female, which in the eagle as in all other
birds of prey is larger than the male, and alfo
feems in the ftate of nature to be bolder, more
intrepid, and more fubtle, appears to lofe its
courage and fagacity when reduced to captivity. The males are preferred for fport ;
and it is obferved that, in the fpring, when the
feafon of love returns, they endeavour to efcape
to their females. And if we employ them dur-

* To this fpecies we may refer the following paffage : " There
are eagles alfo in the mountains near Tauris, in Perfia ; I have feen
one fold for five halfpence by the peafants. People of rank chafe
this bird with the fparrow-hawk. This fport is fomewhat curious and
extraordinary. The fparrow-hawk flies high above the eagle, darts
rapidly upon him, fixes its talons in his fides, and continuing to fly
beats his head with its wings. It fometimes happens that both fall
together.

ing

ing this critical period, we run a rifk of lofing them, unlefs we cool the ardour of their paffion by adminiftering violent purges. It has alfo been remarked, that when an eagle, after leaving the hand, fkims along the ground, and afterwards rifes perpendicularly, he meditates an efcape. He muft inftantly be folicited to return, by throwing him food. But if flies wheeling above his keeper, and does not ftretch to a diftance, it is a fign of his attachment and conftancy. It has been obferved likewife, that an eagle trained for fport, lofing its original inftinct, often attacks and devours the gos-hawk and other fmall birds of prey ; but in the ftate of nature, it only contends with them, or plunders them, as rivals.

In the ftate of nature, the eagle never engages in a folitary chace but when the female is confined to her eggs or her young. This is the feafon when the return of the birds affords plenty of prey, and he can with eafe provide for the fuftenance of himfelf and that of his mate. At other times, they unite their exertions, and they are always feen clofe together, or at a fhort diftance from each other. The inhabitants of the mountains, who have an opportunity of obferving their manœuvres, pretend, that the one beats the bufhes, while the other, perched on a tree or a rock, watches the efcape of the prey. Often they foar beyond the reach of human fight, and notwithftanding the

3   immenfe

immenfe diftance, their cry is ftill heard, and
then refembles the barking of a fmall dog.
Though a voracious bird, the eagle, efpecially
in captivity and deprived of exercife, can
endure for a long time the want of fuftenance. I have been informed by a man of
veracity, that a common eagle caught in a fox-
trap, paffed five whole weeks without the leaft
food, and that it did not appear fenfibly weakened till towards the laft week, after which
they killed it, to put an end to its lingering
pain.

Though the eagles in general prefer defert
and mountainous tracts, they are feldom found
in narrow peninfulas, or in iflands of fmall
extent. They inhabit the interior country in
both continents, becaufe iflands are commonly
not fo well ftocked with animals. The antients
remarked that eagles were never feen in the ifle
of Rhodes, and confidered it as a prodigy, that
when the Emperor Tiberius vifited that famous
fpot, an eagle perched upon the houfe where he
lodged. Eagles make excurfions into iflands,
but do not fix their refidence there, or lay their
eggs ; and when travellers fpeak of eagles,
whofe nefts they find on the fea-fhore or in
iflands, they mean not thofe which we have
mentioned, but the Ofpreys, commonly termed
Sea-eagles, which are birds of a different inftinct,
and which feed on fifh rather than on game.

I ought

I ought here to relate the anatomical obfervations that have been made on the internal ftructure of eagles; and I cannot draw my information from a better fource than the Memoirs of
thofe Gentlemen of the Academy of Sciences
who diffected two eagles, a male and a female,
of the common fpecies. After remarking, that
the eyes were deep funk; that they were of a
pink colour, with the luftre of the topaz; that
the cornea was arched with a great convexity;
that the ligament was of a bright red, the eye-
lids large, and fufficient to cover the whole eye;
they obferved, with refpect to the interior ftructure, that the tongue was cartilaginous at the tip,
and flefhy in the middle; that the *larynx* was
blunt and not pointed, as in moft of the birds
whofe bill is ftraight; that the *œfophagus* was very
large, and widened below to form the ftomach;
that this ftomach was not a hard gizzard, but pliant and membranous like the *œfophagus*, and only
thicker at the bottom; that thefe two cavities,
both the lower part of the *œfophagus* and that
of the ftomach, were very broad, and fuited to
the voracity of the bird; that the inteftines
were fmall, as in all other animals which feed
on flefh; that there was no *cæcum* in the male,
but in the female there were two pretty broad
ones, more than two inches long; that the liver
was large and of a bright red, the left lobe
larger than the right; that the gall-bladder was
large, and about the fize of a chefnut; that the

kidnies

kidnies were fmall, compared with the other
parts, and with thofe of other birds; that the
male-tefticles were only of the fize of a pea,
and of a yellow flefh colour; and that the *ovarium* and *vagina* of the female were like thofe
of other birds.

[A] Linnæus reckons the Rough-footed Eagle as a variety of the
Falco Gallinarius, becaufe it is fmaller, and its wings more variegated. The character of the fpecies is " That the cere and feet
are yellow; the upper part of the body dufky; the lower, tawny
with dun oval fpots; the tail darkifh and ringed."

The ERNE.

THE Erne tribe appears to me to confift of three varieties : the *Great Erne* *, the *Small Erne* †, and the *White-headed Erne* ‡. The two firft are diftinguifhed only by their fize, and the laft fcarcely differs at all from the firft ; and the fole difcrimination is, that it has more white on its head and neck. Ariftotle ‖ defcribes the fpecies alone, and omits to mention the varieties : he fpeaks indeed only of the Great Erne, for he gives it the epithet of *Hinularia*, which denotes that this bird preys upon fawn, that is, young ftags, deer, and roebucks ; a charaéter that cannot belong to the Small Erne, which is too weak to attack fuch large animals.

* *Le Grand Pygargue*, Buff. *Falco Aíbicilla*, Linn. *Pygargus, Albicilla Hirundinaria*, Bel. & Gefn. *Braunfahler Adler*, Frifch. *White-tailed Eagle*, Will. *Cinereous Eagle*, Penn. & Lath.

† *Le Petit Pygargue*, Buff. *Falco Albicaudus*, Linn. *Aquila Albicilla minor*, Briff. *Erne*, Gefn. *Fawn-killing Eagle*, Charl. *Leffer White-tailed Eagle*, Lath.

‡ *Le Pygargue à tête blanche*, Buff. *Falco Leucocephalus*, Linn. *White-headed Eagle*, Penn. *Bald Eagle*, Cat. & Lath.

‖ " There are feveral kinds of Eagles. One is called *Pygargus*, from its white tail. It haunts the plains, groves, and towns. By fome it is called *Hinnularia*. It alfo reforts to the mountains and forefts. ——The other kinds feldom appear in the plains, or in the groves." ARIST. *Hift. An.*

The difference between the Ernes and the Eagles confifts, Firft, In the want of plumage on the legs; the Eagles are clothed as far as the pounces; but the Ernes are naked in all the lower part. Secondly, In the colour of the bill; in the Eagles, it is of a bluifh black; in the Ernes, it is yellow or white. Thirdly, In the whitenefs of the tail; which circumftance has given rife to the name which the Erne has fometimes received of *White-tailed Eagle*. In fact, the tail is white both in the upper and under fide through its whole length. They differ from the Eagles alfo in their inftincts and habits. They fix their refidence not in deferts, or lofty mountains; they haunt the plains or woods that are near the habitations of men. The Erne appears to fhew, like the Common Eagle, a preference to cold countries. It is found in all the northern kingdoms of Europe. The Great Erne is of the fame fize and the fame ftrength, if not more vigorous, than the Common Eagle : it is at leaft more bloody and ferocious, and lefs attached to its young; for it feeds them but a fhort time, drives them from its neft before they can procure fuftenance ; and it is pretended that, without the affiftance of the Ofprey, which generally takes them under its protection, they would perifh. It has commonly two or three young, and builds its neft upon large trees. A defcription of one of thefe nefts occurs in Willoughby, and in feveral other authors who have

6   copied

copied it. It is an airy, or floor quite flat, like
that of the Great Eagle, sheltered above by the
foliage of trees, and formed with small sticks
and branches, which are covered with several
alternate layers of broom, and other plants.
That unnatural disposition which instigates those
birds to expel their young before their feeble
strength is able to procure an easy subsistence,
and which is common to the Erne, the Golden
Eagle, and the Spotted Rough-footed Eagle,
proves that these three species are more voracious, and more inactive in the pursuit of their
prey, than the Ring-tail Eagle, which watches
and feeds* generously its infant brood, and afterwards trains them, teaches them to hunt, and
does not desert them till their dexterity and vigour are sufficient for their support. The young
also inherit the instinct of their parents. The
Eaglets of the common kind are gentle and
peaceful; but those of the Golden Eagle and
the Erne, as soon as they have acquired some
stature, are continually fighting and contending
about their food, and their place in the nest:
so that the father and mother, to terminate the
quarrel, often destroy a mutinous subject. The
Golden Eagle and the Erne generally point

---

* " The ossifraga feeds carefully both its own young and those
of the eagle; for when it ejects them from the nest, this bird receives them, and breeds them. The eagle turns them out before
they can procure food, or fly. In this forlorn state, the ossifraga
listens to their complaints, and kindly takes them under its protection." ARIST. *Hist. An.*

their

their attacks upon large animals; they often
fatiate themfelves upon the fpot, being unable
to tranfport their prey: hence their depre-
dations are lefs frequent, and, not preferving
carrion in their neft, they are often reduced
duced to want. On the other hand, the Common Eagle, which catches every day hares and
birds, fupplies more eafily and more plentifully
the neceffary fubfiftence to its young. It has
alfo been remarked, efpecially with regard to the
Ernes, which chufe their haunt near fettled
fpots, that they fearch for their prey only during a few hours in the middle of the day, and
devote the morning, the evening, and the night
to fleep; whereas the Common Eagle *(Aquila
Valeria)* is more adventurous, more active, and
more indefatigable [A].

[A] The three birds claffed together in this article are confidered by Linnæus, and other fyftematic writers, as diftinct fpecies:—
  Firft, The Great Erne, or Cinereous Eagle; *Falco albicilla,* LINN.
Its character is, " That its cere and feet are yellow; the tail-feathers white, and the intermediate ones tipt with black." It is of
the fize of a peacock, being two feet nine inches long; its head and
neck are of a pale afh-colour; the iris and bill pale yellow, and the bill
elongated at its bafe; the fpace between the eyes and the ears naked, with fmall ftraggling briftles, and of a cærulean hue. The body
and wings are cinereous, with dun intermixed; the tail white; the
feet woolly below the knees, and of a bright yellow; the claws
black.——It inhabits Europe, particularly Scotland and the adja-
cent iflands, and preys upon large fifh.
  Second, The Little Erne, or White-tailed Eagle; *Falco albicandus,*
LINN.—" Its cere and feet are naked and yellowifh; the head and
neck afh-coloured, bordering on chefnut; the body of a dull ferruginous above, and below ferruginous and blackifh; the tail white."
It is of the bulk of a large cock, being two feet two inches long.
The

The bill and iris are inclined to yellow; the tips of the quill-feathers verging on black; the nails black.

Third, The White-headed Erne, or Bald Eagle; *Falco leucocephalus*, Linn.—" The cere yellowish; the feet partly woolly; the body dusky; the head and tail white." It is three feet three inches long, and weighs nine pounds. Its head grows white till the second year. It preys on fawns, pigs, lambs, and fish. It watches the motion of the Osprey; and as soon as that bird has seized a fish, it pursues till the prey drops, and, with astonishing dexterity, catches it before it falls to the ground. It builds in the forests of maples, cypresses, and pines, generally on the margin of water; and its nests are so much crowded as to resemble a rookery. They are very large, and have a stench from the fragments of carrion. In Bering's Island, the Bald Eagle nestles on the cliffs, and lays two eggs in the beginning of July.

The OSPREY.

Le *Balbuzard*, Buff.
Falco Haliaëtus, Linn.
The Bald Buzzard, Will
The Morphnos, or *Clanga*, Ray and Will.
Fishaar. Wires.
In Italian, *Anguista Piombina*.
In Polish, *Orzelmarsky*.

IF we consider all the facts relating to this
bird *, we must conclude that, though it resembles the eagle more than any other bird of
prey, it really constitutes a distinct genus †. It
is much smaller, and has neither the port, the

* In the beginning of this article Buffon remarks, that the
Osprey is often called the *Sea-Eagle*; and that in Burgundy it goes
by the name of *crau pêcherot*, or *crow-fisher*; a word which, as well
as many more, has been introduced into the dialect of the peasants
of that province by the residence of the English troops.

† The difference between the male and female of the Osprey
is still greater than in the eagles. The one described by Brisson,
which was undoubtedly a male, was only one foot seven inches
long, measured to the nails, and five feet three inches across
the wings; and another which was brought to me, was one foot
nine inches in length, and five feet seven inches of alar extent.
But the female, described by the anatomists of the Academy of
Sciences by the name of *haliætus*, was two feet nine inches long,
including the tail, which would allow at least two feet for the body;
and the alar extent was seven feet and a half. This difference is
indeed so great, that, were it not for other indications, we should
doubt if the bird described was really the *Osprey* or *Bald-
Buzzard*.

figure,

THE OSPREY.

figure, nor the flight of the eagle. Its natural
habits are as different as its appetites; for it
feeds chiefly on fifh, which it catches in the
water, and even feveral feet below the furface *.
And that this is its ordinary fubfiftence appears,
becaufe its flefh has a ftrong fifhy flavour. I
fometimes obferved this bird remain more than
an hour perched upon a tree contiguous to
a pool, watching the appearance of a large fifh,
and ready to dart upon it and tranfport the
victim in its talons. Its legs are naked and
commonly of a bluifh colour; in fome individuals, however, the legs and feet are yellowifh,
the claws large and fharp, the feet and toes
fo ftiff that they cannot be bent; the belly is
entirely white, the tail broad, and the head thick
and bulky. It differs from the eagles, becaufe
its feet and the lower part of its legs are not
feathered, and its hind pounce is the fhorteft;

* Ariftotle, however, ranks the Ofprey among the eagles :—
" The fifth fpecies of eagles is the *Haliætus* or *Sea-Eagle*; its
" neck large and thick, its wings curved, its tail broad. It
" haunts the fhores. When it is unable to carry its prey, it often
" plunges it into the water." We muft obferve that the Greeks
claffed the rapacious birds into three kinds; the Eagle, the Vulture,
and the Hawk. This explains the arrangement which Ariftotle
has here adopted. But it is fomewhat fingular, that Ray, who is
otherwife an intelligent and accurate writer, fhould affirm that the
Offifraga and the *Haliætus* are the fame; fince Ariftotle diftinguifhes them, and treats of them in two feparate chapters. The
only reafon which Ray could urge, is the fmallnefs of the Ofprey or
Bald-Buzzard ; but the fame reafon would alfo exclude the *Morph-
nus* from the eagles.

while

while in the eagles, that is the longeſt of all.
It is diſtinguiſhed from the eagle by another
circumſtance; for its bill is of a deeper black,
and the feet, the toes, and the cere are commonly
blue, while thoſe of the eagle are yellow. The
toes of the left foot are not connected by ſemi-
membranes, as Linnæus aſſerts *; for the toes
of both legs are alike parted and devoid of
membranes. It is a popular error, that this bird
ſwims with the one foot, and catches fiſh with
the other; and this has occaſioned the miſtake of
Linnæus. Formerly Klein affirmed the ſame
thing of the Great Sea-eagle, but he was equally
miſtaken; for neither of theſe birds has membranes connecting the toes of the left foot. The
common ſource of theſe errors is Albertus Magnus, who writes, that this bird had one foot like
that of the ſparrow-hawk, and the other reſembling that of the gooſe; an aſſertion which
is not only falſe, but abſurd and inconſiſtent
with every analogy. It is indeed aſtoniſhing,
that Geſner, Aldrovandus, Klein, and Linnæus,
inſtead of rejecting this ſilly fable, have blindly
adopted it; and that Aldrovandus tells us coolly,
that it is not improbable, ſince he poſitively
adds, he knows there are ſeveral water-fowls
whoſe feet are half-cloven, half-webbed; an aſſertion as falſe as the firſt.

* Linnæus has omitted this account in the later editions [a].

I am

I am not furprifed that Ariftotle called this bird
ἀλιαέτος, Sea-Eagle ; but I am aftonifhed that all
the naturalifts, ancient or modern, have copied
the name without fcruple, and I might fay
without reflexion; for this bird frequents the
fea-fhore not from any decided preference. It
oftener haunts inland countries that are contiguous to rivers, lakes, and other frefh waters;
and it is even more common in Burgundy, which
is the centre of France, than on any of our
coafts. As Greece is a country which has few
rivers, pools, or lakes, and which is much interfected and indented by the incroachment or
retreat of the ocean, Ariftotle obferved that thefe
bird-fifhers fought their prey on the beach, and
for this reafon he named them *Sea-Eagles*. But
had this philofopher lived in the heart of France,
Germany *, Switzerland †, or any other country diftant from the fea; and where thefe birds
are common, he would probably have termed
them *frefh-water-eagles*. Ariftotle affirms, that
this bird has a keen fight; it compels its young,
he fays, to look at the fun, and kills thofe whofe
eyes cannot fupport the glare. I cannot authen-
ticate this fact, which appears to me rather

---

* " This bird is exceedingly fat, and is entirely of a fifhy odor.
" It lives not only on the fea-coaft, but haunts the rivers and lakes
" in Silefia, and fits in the trees watching the fifh." SCHWENCK-
FELD.

† Gefner fays, that this bird is found in feveral parts of Switzerland, and makes its neft in certain rocks near waters, or in deep
vallies. He adds, that it can be tamed, and employed in falconry.

improbable,

improbable, though it has been related or rather repeated by several authors; and though it has even been generalized and attributed to all the eagles, which are said to force their brood to look steadily at the sun. This is an observation which it would be difficult to make; and Aristotle was besides not much acquainted with the facts relating to the young of this bird. He alleges that they rear only two, and kill that one whose eyes cannot bear the dazzling rays of the sun. But we are certain that they often lay four eggs, and seldom three only, and that they raise all which are hatched. Instead of inhabiting rocky precipices and lofty mountains, as do the eagles, it prefers the haunts of low and marshy grounds, in the vicinity of pools and lakes that abound with fish. It appears that we must ascribe to the *Osprey*, and not to the *Sea-Eagle*, what Aristotle mentions with regard to the pursuit of sea-birds; for the Sea-eagle fishes rather than hunts; nor have I heard that it strays to a distance from the beach in the chace of gulls and other sea-birds: on the contrary, it seems to subsist entirely upon fish. Those who have dissected this bird, have found nothing but fish in its stomach; and its flesh, which, as I have said, has a very strong smell of fish, is a sufficient proof that this constitutes its ordinary food. It is commonly very fat, and can, like the eagles, support for several days the want of sustenance, without suffering inconvenience or loss of strength. It

is

is not fo bold or fo ferocious as the eagle or erne ; and it is pretended that it could be as eafily in-ftructed for fifhing, as the other birds are trained for the fport.

After comparing authorities, I am of opinion that this fpecies is one of the moft numerous of the large birds of prey ; and is fcattered over the extent of Europe, from Sweden to Greece ; and that it is even found in warmer countries, as in Egypt and Nigritia.

The internal parts of this bird differ little from thofe of the eagles. The academicians have per-ceived the moft confiderable diftinction in the liver, which in the Sea-eagle is very fmall. The *cæcum* of the female alfo is not fo large ; and the fpleen, which in the eagle is clofely attached to the right fide of the ftomach, is in this bird placed under the right lobe of the liver. Like thofe of moft other birds, its kidneys are proportionally large, whereas thofe of the eagle are fmall [A].

[A] The fpecific character which Linnæus gives of the Ofprey, *Falco Haliaëtos*, is, " that its cere and feet are cœrulean, its body " dufky above and white below, and its head whitifh." Befides the *Carolina Ofprey*, which is ranged by Buffon among the foreign analogous birds, there are two other varieties : 1. The *Reed Ofprey, Falco Arundinaceus*, GMEL.—" Its cere is afh-coloured, its " feet pale, its body grey above and whitifh below, and its tail " is equal." 2. The *Cayenne Ofprey, Falco Cayanenfis*, GMEL.— " Its body is dufky ferruginous, and there is a white line drawn " from the upper mandible through each eye to the hind part of " the head, which is alfo white."

The SEA-EAGLE.

L'Orfraie, Buff.
Falco Offifragus, Linn.
Aquila Offifraga, Briff. and Klein.
In Italian, *Aquilaftro anguifta barbata.*
In German, *Groffer hafenabr.*
In Polifh, *Orzel Lomignat.*

THIS bird has been called by our nomenclators, the *Great Sea-Eagle.* It is indeed
nearly as large as the Golden Eagle; and its
body feems proportionally longer, though its
wings be fhorter. It meafures, from the end of
the bill to the point of the nails, three feet and
an half; but its expanded wings do not reach
above feven feet: whereas the length of the
Golden Eagle is generally only three feet two
or three inches, and the extenfion of its wings
eight or nine feet. The Sea-Eagle is remarkable for its fize, and is diftinguifhed: 1. By the
colour and figure of its nails, which are of a
fhining black, and form an entire femicircle.
2. By its legs, which are naked below, and
covered with fmall yellow fcales. 3. By the
beard of feathers which hangs from the chin,
and which has occafioned its receiving the name
of *Bearded-Eagle.* The Sea-Eagle loves to
haunt the fea-fhore, and often frequents inland
tracts,

THE SEA EAGLE

tracts, near lakes, marfhes, or rivers that are ftocked
with fifh. It catches the largeft of the finny tribe;
but it alfo attacks game, and, as it is large and
ftrong, it feizes and carries off geefe and hares,
and even lambs and kids. Ariftotle affures us,
that the female Sea-Eagle not only watches
her infant brood with the greateft affection, but
extends her protection to the young eaglets,
which have been expelled by their unfeeling
parents, and generally feeds and trains them as
if they were her own offspring. This fingular
fact has been repeated by all the naturalifts, but
it does not appear to be authenticated. We
have reafon however to admire the general accuracy of his Hiftory of Animals. We have a
remarkable proof of this in point:—Ariftotle
obferves, that the fight of the Sea-Eagle is weak,
on account of a fhade which covers the eye;
and hence he was probably induced to feparate
it from the eagles, and clafs it with the owls
and night birds. To judge of this fact by the
confequences, we fhould infer, that it is not only
doubtful, but falfe; for all who have watched
the manœuvres of the Sea-Eagle, have found
that it fees during the night fo diftinctly as to
be able to catch game, and even fifh; but they
have not obferved that its fight was feeble during
the day. On the contrary, it perceives at a
great diftance, the fifh upon which it is to dart;
it purfues with eagernefs the birds on which it
is to prey; and though it flies with lefs rapidity
than

than the eagles, this is owing to the fhortnefs of its wings, and not the indiftinctnefs of its vi-fion. A refpect for the great philofopher of antiquity has induced the celebrated Aldrovandus to examine the eye of the Sea-Eagle with mi-nute attention; and he has difcovered that the aperture of the pupil, which is commonly cover-ed only with the cornea, is in this bird lined befides with an exceedingly delicate membrane, which has actually the appearance of a fmall fpot. He alfo obferved, that the inconvenience of this ftructure is compenfated by the perfect tranfparency of the circular part furrounding the pupil, which ring in other birds is opaque and of a dull colour. Thus the obfervation of Ariftotle is good, fince he has remarked that the eye of the Sea-Eagle is covered with a thin cloud; but it does not neceffarily follow, that its fight is fainter than that of other birds, becaufe the light can pafs eafily and largely through the fmall circle which bounds the pupil; all that can be inferred is, that the middle of the pic-ture muft be marked with a fmall obfcure fpot, and that the lateral vifion ought to be more diftinct than the direct. It is not however mani-feft from what has been faid, that it fees worfe than the other birds. It foars not indeed to the fame height as the eagle, nor flies with equal rapidity; nor does it defcry or purfue its prey from fo remote diftances. It is probable, there-fore, that the fight of the Ofprey is not fo acute

or

or diftinct as that of the eagle; at the fame time
it is not, like the owls, blinded by the dazzling
light, but fearches for victims in the day as well
as in the night *, particularly in the mornings
and evenings. Befides, if we compare the eye
of the Sea-Eagle with that of the owl and
other nocturnal birds, we fhall perceive a difference of ftructure. In the day-time thefe
birds fee faintly, if at all; their delicate organs
being unable to bear the fhock of a blaze of
light; their pupil is entirely open, and not lined
with that membrane or fmall fpot, which is found
in the eye of the Sea-Eagle. The pupil in all
nocturnal birds, cats, and fome other quadrupeds,
which fee in the dark, is round and large, when
it receives the impreffion of a faint light in the
dufk of the evening. On the contrary, it becomes elongated in cats, but in the nocturnal
birds, it retains its globular figure, though it
contracts its fize, when the eye is expofed to a
ftrong glare. This contraction evidently proves
that thefe birds fee *ill*, becaufe they fee too *well*;
fince their delicate organs are fenfible to the
fainteft impreffion. Wherefore, *à fortiori*, the
Sea-Eagle, with its fpot upon the pupil, would
require more light than any other, if this de-

---

* I have been informed by perfons who obferved the fact, that
the Sea-Eagle catches fifh during the night, and that the noife of
its plunging into the water is heard at a great diftance. Salerne
alfo remarks, that when it darts into a pool to feize its prey, the
noife it occafions is terrible, efpecially in the night.

fect

fect were not compensated. But what forms a
complete apology for Ariftotle's arranging it
among the nocturnal birds is, that the fight of
the Sea-eagle is not fo acute as that of the Common Eagle; and becaufe it commits its ravages by
night as well as by day.

If the facts related by Ariftotle in his Hiftory
of Animals, be diftinguifhed by their accuracy,
his treatife *De Mirabilibus* is no lefs remarkable for its abfurdities and errors. The author
even makes affertions in it which are totally inconfiftent with what he has delivered in his other
works; and if we were to compare the opinions,
but particularly the facts, with thofe in his Hiftory of Animals, we never fhould afcribe the
treatife *De Mirabilibus* to that enlightened philofopher. Pliny, whofe Natural Hiftory is entirely extracted from Ariftotle, would not have
related fo many things that are falfe or equivocal,
had he not borrowed indifcriminately from the
different treatifes attributed to the Greek, and
collected the opinions of fubfequent authors
which are tinctured conftantly with popular prejudices. We can give an example without deviating from our prefent fubject. Ariftotle diftinguifhes the fpecies of the Ofprey in his
Hiftory of Animals, fince he makes it the fifth
fpecies of eagles, to which he gives accurate
difcriminating characters. At the fame time,
the Ofprey conftitutes not a diftinct fpecies in
the treatife *De Mirabilibus*, or rather is only a
variety.

variety. Pliny, enlarging on this idea, affirms
not only that the Ofpreys form no feparate race,
and that they proceed from the intermixture of
the different fpecies of eagles, but that the young
are not Ofpreys, and only Sea-Eagles; *which
Sea-Eagles*, fays he, *breed fmall vultures, which
engender great vultures, that have not the power
of propagation.* What a number of incredible
circumftances are grouped into this paffage?
How many things that are abfurd, and contrary
to every analogy? Let us even extend, as much
as poffible, the probable limits of the variations
of Nature, and let us give this paffage the moft
favourable explanation. Suppofe for a moment,
that the Ofpreys are really the hybriduous off-
fpring of the union of two different fpecies of
eagles; that they are prolific, like the crofs-breed
of fome other birds, and produce between them
a fecond mongrel, which approaches nearer the
fpecies of Sea-Eagle than if the firft mixture
were that of the Sea-Eagle with another eagle;
fo far the laws of Nature are not entirely violated: but to add, that thefe Ofpreys, after
they become Sea-Eagles, breed fmall vultures,
and thefe again the great, which are incapable
of generation, is to join three facts that are
abfolutely incredible, to two that already can
hardly be believed. And though Pliny has
written many things haftily, I can hardly per-
fuade myfelf that he is the author of thefe three
affertions; and I am rather inclined to fuppofe,

that the end of the fentence has been entirely altered. At any rate it is certain that the Sea-Eagles never breed fmall mongrel vultures, nor do thefe give birth to large hybriduous vultures, whofe prolific powers are extinguifhed. Every fpecies of vulture produces its like, and the fame is the cafe with each of the eagles, the Ofprey and the Sea-Eagle; and the intermediate kinds, bred by the intermixture of the eagles, confti-tute independent tribes, and are perpetuated like the others by a generation. Particularly, we are well informed, that the male Ofprey breeds with its female young Ofpreys; and that, if it ever begets Sea-Eagles, this is only in the union with the Sea-Eagle. The copulation of the male Ofprey with the female Sea-Eagle is fimilar to that of the he-goat with the ewe: a lamb is the fruit of this commerce, becaufe the influence of the ewe predominates in the conception. A Sea-Eagle is alfo the product of the other inter-courfe; for, in general, the character of the female preponderates, and the prolific mongrels approach to the fpecies of the mother; and even the true hybrids, or the barren mongrels, bear a greater refemblance to the race of the female than to that of the male.

What renders the poffibility of the crofs-breed of the Sea-Eagle and the Ofprey credible, is the fimilarity of the inftincts, the difpofitions, and even the figure of thefe birds; for though
they

they differ widely in point of fize, the Sea-
Eagle being near one half larger than the Ofprey,
they are yet very much alike in their proportions. Their wings and legs are fhort, compared with the length of their body; the lower
part of their legs and feet are naked; they fly
neither fo high nor fo rapidly as the eagles; they
derive their fubfiftence more from the finny
tribe than from the beafts of game; and they
haunt places contiguous to lakes or fifhy ftreams;
and both of them are common in France and
other temperate countries. The Sea-Eagle, as
it is larger, lays only two eggs, and the Ofprey
four *. The cere of the Ofprey and the legs
and feet are generally blue, but the fame parts
of the Sea-Eagle are of a bright deep yellow.
There is alfo fome difference in the diftribution
of the colours of the plumage, but this is too
flight to prevent their intermixture; and ana-

* The Sea-Eagle builds on the loftieft oaks a very broad neft,
into which it drops two very large eggs, that are quite round, exceedingly heavy, and of a dirty white colour. Some years ago
one of thefe was found in the park of Chambord; I fent the two
eggs to Reaumur, but the neft could not be detached. Laft year
a neft was difcovered at St. Laurent-des-eaux, in the wood of Briou,
but there was only one eaglet, which the poft-mafter of that place
has raifed. At Bellegarde, in the foreft of Orleans, a Sea-Eagle
was killed, which during the night fifhed the largeft pikes in a
pond that belonged formerly to the Duc d'Antin. Another was
killed lately at Seneley in Sologne, in the moment it was carrying
off a large carp in broad day. The Ofprey lodges among the
reeds, along the margin of waters; it lays each hatch four white
eggs of an elliptical fhape: it feeds on fifh. SALERNE.

logical

logical reafons induce me to fuppofe that the
union is prolific, and that the male Ofprey, by
coupling with the female Sea-Eagle, produces
Sea-Eagles; but that the female Ofprey, by
pairing with the male Sea-Eagle, gives birth to
Ofpreys; and that fuch hybrids, whether they
be Sea-Eagles or Ofpreys, inherit almoft entirely
the nature of their mother, and retain but flight
traces of the character of the father; which circumftance diftinguifhes them from legitimate
Ofpreys or Sea Eagles. For example; Ofpreys
fometimes occur with yellow feet, and Sea-Eagles
with blue feet, though the reverfe commonly
takes place. This variation of colour muft arife
from the mixture of the two fpecies. For the
fame reafon, Ofpreys are found, fuch as what
the members of the Academy have defcribed,
that are much larger than ordinary; and at the
fame time fome Sea-Eagles are much fmaller
than common; and the diminutive fize of thefe
can be afcribed neither to the fex nor to the age,
but muft arife from a mixture with the fmaller
fpecies; that is, of the Ofprey with the Sea-
Eagle.

As this bird is very large, and confequently
little prolific, laying only two eggs once a year,
and often raifing but a fingle young one, the
fpecies is no where numerous. It is however
widely diffufed; it is found in almoft every part

of

of Europe, and it even appears to frequent the lakes of North America*. [A]

* I conceive that the following passage alludes to the Sea-Eagle;
" There are a number of eagles, which in their language are called
fondaqua. They commonly build their nests on the margin of
water, or of some precipice which overtops the highest trees or
rocks, so that it is difficult to reach their nests. We have however
got several, but never found more than two eaglets. I intended
to breed some when we were on the road from Lake Huron to
Quebec; but they being heavy to carry, and we not being able to
give them fish to feed on, we dressed them into an excellent dish,
for they were then young and tender." *Voyage au Pays des Hurons
par Sagar Theodat.*

[A] The Sea-Eagle, *Falco ossifragus,* LINN. is thus described:
" Its cere is yellowish, its feet partly woolly, its body ferrugi-
" nous, the tail-feathers white along the inside." It resembles
the Golden Eagle, and is of the size of a turkey. It is sometimes
drowned in attempting to catch overgrown fish; not being able to
disengage its talons, it is dragged forcibly under water. The Tartars entertain a notion that the wound of its claws is mortal, and
therefore they dread its attack.

The WHITE JOHN*.

Jean le Blanc, Buff. *and* Lath.
Falco Gallicus, Linn.
Falco Hypoleucos, Decouv. Ruff.
Aquila Pygargus, Briff. Johnft. Belon.
L'Albanella, Let. uc Sard.
Blanche-Queue, Hift. de Lyon.

I HAD this bird alive, and kept it for some time. It was taken young in the month of Auguft 1768, and it appeared in January 1769 to have attained its full fize. Its length, from the point of the bill to the extremity of the tail, was two feet; and to the pounces one foot eight inches: the bill from the hook to the junction of the mandible feventeen lines; the tail was ten inches long; its wings when expanded meafured five feet one inch, and when clofed they reached a little beyond the end of the tail. The head, the upper furface of the neck, the back, and the rump, were of an afh-brown; all the feathers which cover thefe parts were white at their origin, but the reft of them brown; fo that the brown concealed the white, except the

---

* " Some have called it *The White-tailed Knight*, becaufe perhaps it is rather tall.——The male is lighter and whiter than the female, particularly on the rump; the tail very long, the legs flender, and of a pleafant yellow."—SALERNE.

plumage

THE WHITE JOHN.

plumage was raifed. The throat, the breaft, the belly, and the fides were white, variegated with long fpots of a brown rufty colour; there were tranfverfe bars of a deeper brown upon the tail; the cere was of a palé blue, and in it were placed the noftrils. The iris was of a beautiful yellow citron, or of the colour of the oriental topaz; the feet were of a pale flefh hue, which, as well as the cere, paffed into yellow as the bird grew older. The interftices of the fcales which covered the fkin of the legs, appeared reddifh; fo that the whole, feen at a diftance, appeared, even in its tender age, to be yellow. This bird weighed three pounds feven ounces after a meal, and three pounds four ounces when its ftomach was empty.

The *White John* is more widely removed from the eagles than any of the preceding. It refembles the *Bald Eagle* only by the want of plumage on the legs, and the whitenefs of the rump and tail; but its body is differently fhaped, and is much thicker, compared with its bulk; for its extreme length is only two feet, and the expanfion of its wings feven feet; while the girth of its body is as great as that of the Ring-tail Eagle, the length of which is two feet and an half, and the alar extent more than feven. Hence the *White John* refembles in its fhape the Ofprey, whofe wings are alfo fhort in proportion to its body; but its feet are not, as in that

bird, blue : its legs alſo appear to be more ſlender and tapered than any of the Eagles.. Thus,
though it bears ſome analogy to the Eagles, the
Oſprey, and the Sea-eagle, it is yet quite a diſtinct ſpecies. It has ſome reſemblance alſo to
the Buzzard in the diſpoſition of its colours, of
its plumage, and alſo in a circumſtance which
has often ſtruck me, viz. that, in certain attitudes, and eſpecially in the front view, it appears like an Eagle ; and that, ſeen ſideways or
in other attitudes, its figure is ſimilar to that of
the Buzzard. This remark was alſo made by
my deſigner and others ; and it is ſingular that'
this ambiguity of figure correſponds to its equivocal diſpoſition, which is really analogous both
to that of the Eagle and that of the Buzzard ;
inſomuch that the White John may, in certain
reſpects, be conſidered as forming the intermediate ſhade between theſe two birds.

It appeared to me that this bird ſaw very
diſtinctly in the day-time, and was not afraid of
the ſtrongeſt light ; for it ſpontaneouſly directed
its eyes to the moſt luminous quarter, and even
to the ſolar effulgence. It ran with conſiderable
ſwiftneſs when ſcared, and aſſiſted its motion by
its wings. When confined to its chamber, it
ſought to approach the fire ; but cold did not
ſeem to be abſolutely pernicious to it, for it
paſſed ſeveral nights in open air in froſty weather without appearing to ſuffer inconvenience.

It

It fed upon raw bloody flesh ; but, when pinched with hunger, it ate meat that had been cooked. It tore the flesh that was offered it with its bill, and swallowed it in large morsels. It never drank when any person was beside it, or was within its sight ; but when it was in a concealed place, it was observed to drink, and to use more precaution than might be expected. A vessel filled with water was left within its reach ; it looked anxiously on every side, to ascertain that it was quite alone ; it then approached the vessel, but still cast an attentive look around : at last, after many hesitations, it plunged its bill up to the eyes in the water, and repeated its draught. It is probable that other birds of prey conceal themselves in the same manner when they want to drink ; the reason probably is, that these birds can take no liquid but by immersing their head beyond the opening of the mandibles, and even as far as the eyes ; in which case, they are thrown off their guard, and have reason to entertain fears : however, this is the only circumstance in which the White John shewed any mistrust ; and in other things he appeared indifferent, or rather stupid. He was not at all mischievous, and suffered himself to be handled without discovering resentment. He uttered the sound, co, co, a slight expression of contentment when food was offered him ; but he shewed no particular attachment to any individual. He grew fat in the autumn, and got

more

more flesh, and became plumper than most other birds of prey *.

This bird is very common in France; and, as Belon says, there is hardly a cottager who is not acquainted with it, and who dreads not its ravages among his poultry. The peasants have given it the name of *White John* †, because it is remarkable for the whiteness of its belly, of the under surface of its wings, of its rump, and of its tail : these characters, however, are distinctly

* The following note respecting this bird was given to me by the person entrusted with my volaries :—" I offered the White John several kinds of food ; as bread, cheese, grapes, apples, &c. but he would taste none of them, though he had fasted twenty-four hours. I kept him in that state three days longer, and he still rejected them. In short, we may conclude, that hunger will never drive him to eat any thing of that sort. I gave him worms, but he constantly rejected them ; and though one was forced into his bill, and half swallowed, he threw it out. He greedily swallowed mice entire ; and I observed, that after swallowing two or three, or even a single large one, he shewed uneasiness, and marks of pain, drooping and inactive, and stupid with digestion. I presented frogs and small fish ; the latter he rejected, but ate the frogs by half-dozens, and sometimes more ; but he did not swallow them whole as he did the mice ; he fixed his talons, and tore them into pieces. I withheld food from him three days, and he still refused raw fish. I observed that he vomited the skins of the mice in little balls of an inch, and I found, by steeping them in water, that they contained only the hair and the skin, and not the bones : in some of these balls were particles of cast iron, and bits of coal."

† " The inhabitants of the villages know this bird to their cost ; for it plunders their poultry more daringly than the kite." BE-LON.

" This White John attacks the hens in the villages, and catches birds and rabbits. It is very destructive to partridges, and preys on small birds ; for it plunders, concealed along the hedges, and the skirts of the forests ; so that not a peasant is unacquainted with it." *Id.*

8   marked

marked only in the male; for the female is almoſt entirely grey, and the feathers of the rump
alone of a dirty white. As in the other birds
of prey, ſhe is larger than her mate; ſhe neſtles
almoſt cloſe upon the ground, in tracts covered
with heath, fern, broom, or ruſhes; ſometimes,
however, ſhe builds on firs, and other high trees.
She commonly lays three eggs, which are of a
grey ſlate-colour; the male provides largely for
her ſubſiſtence during the time of incubation,
and even while ſhe is employed in watching
and educating her young. He haunts the vicinity of inhabited places, eſpecially near hamlets
and farms; he plunders chickens, young turkies, and tame ducks; and when poultry cannot be had, he catches young rabbits, partridges, quails, and other ſmall birds; nor does
he diſdain the more humble prey of field-mice
and lizards. As this bird, particularly the female, has ſhort wings and a thick body, the
flight is laborious, and they never riſe to a great
height; they conſtantly ſkim along the ground,
and commit their ravages upon the earth rather
than in the air *. Their cry is a kind of ſhrill
whiſtling, which is ſeldom heard. They ſcarcely ever ſeek their prey but in the morning and
the evening; and the middle of the day is de-
voted to indolence and repoſe.

* " When we obſerve it flying, it appears like a heron; for it
ruſtles with its wings, and does not ſoar like many other birds of
prey, but generally flies near the ground, eſpecially in the morning and evening."—BELON.

One

One fhould be apt to fuppofe that there is a variety in this fpecies; for Belon gives a defcription of a fecond bird, " which is," he fays, " another fpecies of the St. Martin, alfo named " *White-tail*, of the fame fpecies with the above- " mentioned White John, and which refembles " the Royal Kite fo exactly that we could dif- " cover no difference, except that it is fmaller, " and whiter under the belly, the feathers of its " rump and tail being on both fides of a white " colour." Thefe points of refemblance, to which we may add what is ftill more important, that its legs are longer, prove only that this fpecies is allied to that of the White John: but as it differs confiderably in its fize, and in other circumftances, we can but infer that it is a variety of the White John; and we have perceived that it is the fame bird which our nomenclators have called the *Cinereous Lanner;* and which we fhall mention under the name of St. Martin, becaufe it has not the leaft refemblance to the *Lanner.*

The White John, though very common in France, is unfrequent in every other country; fince none of the naturalifts of Italy, England, Germany, or the North, mention it, except from the authority of Belon. For this reafon I have dwelt more fully upon the facts relating to its hiftory. I muft alfo obferve, that Salerne commits a great miftake, when he fays that this bird is the fame with the *Ringtail* of the Eng-

lifh,

lifh, the male of which is termed *Hen-harrier*. The character of the White-tail, and the proneness to prey on poultry, common to the Ringtail and the White John, have deceived him, and induced him to confider thefe birds as the fame ; but if he had compared the defcriptions of preceding authors, he would have eafily perceived that they belong to different fpecies. Other naturalifts have taken the *Blue-hawk* of Edwards for the *Hen-harrier*, though thefe birds are alfo of different kinds. We fhall endeavour to clear up this point, which is one of the moft obfcure in the natural hiftory of the rapacious tribe.

Birds of prey are divided into two orders : the firft of which includes the warlike, the noble, and the intrepid ; fuch as Eagles, Falcons, Ger-falcons, Gos-hawks, Lanners, Sparrow-hawks, &c. : the fecond comprehends thofe that are indolent, cowardly, and voracious ; fuch as, the Vultures, the Kites, the Buzzards, &c. Between thefe two orders, fo oppofite in their inftincts and habits, are found, as every where elfe, fome intermediate fhades, or fome fpecies which participate of the character of both. Thefe are : Firft, The White John, of which we have now treated, and which, as we have faid, is a-kin to the Eagle and the Buzzard. Secondly, The St. Martin, which Briffon and Frifch have called the *Cinereous Lanner*, and Edwards has named the *Blue Falcon*, but which refembles

more

more the White John and the Buzzard than the Falcon or Lanner. Thirdly, The *Sou-buse*, with which species the English seem not to have been well acquainted, having taken another bird for the male, whose female they have named *Ring-tail*, and the pretended male *Hen-barrier*. These are the birds which Brisson has called *Collared Falcons*; but they have more affinity with the Buzzard than the Falcon or the Eagle. These three species, and particularly the last, have been misrepresented, or confounded, or improperly named; for the White John ought not to be ranged among the eagles. The St. Martin is neither a Falcon, as Edwards says; nor a Lanner, as Frisch and Brisson assert; for in its instinct it is different, and in its habits it is opposite to those. It is the same with the *Sou-buse*, which is neither an eagle nor a falcon, since its appetites are entirely dissimilar to those of these two species.

But I am of opinion that we ought to class with the White John, with which we are well acquainted, another bird, known only by the indication of Aldrovandus, under the name of *Laniarius*; and of Schwenckfeld, under that of *Milvus albus*. This bird, which Brisson has also called the *Lanner*, appears to me to be more different from the true Lanner than the St. Martin. Aldrovandus describes two of these birds; the one of which is much larger than the other, being two feet from the point of the

bill

bill to the end of the tail, and is of the fize of
the White John; and appears, from comparing
the account of that naturalift, to have the fame
characters.   Nor need we be furprifed that
Aldrovandus, whofe ornithology is on the whole
excellent, efpecially with regard to the European birds, fhould commit this overfight, fince
he derives his acquaintance of the White John
entirely from Belon *, and has even borrowed
his figure.   [A]

* *Pygargi fecundum genus.*   ALDROV.

[A] The White John, *Falco Gallicus*, LINN. The fpecific character :—" The bill cinereous; the feet naked and yellowifh; the
body of a dufky grey, and below (in the male) whitifh, with tawny
fpots."

FOREIGN BIRDS,

RELATED TO THE EAGLES AND OSPREYS.

I.

THE Bird of the East Indies, which Brisson describes accurately, by the name of the *Pondicherry Eagle*. We shall only observe, that its diminutive size alone ought to exclude it from the Eagles, since it is only half the bulk of the smallest. It resembles the Osprey in the colour of the cere, which is bluish; but its feet are not blue as in that bird, nor yellow as in the Erne. Its bill, of an ash-colour at its origin, and of a pale yellow at the tip, seems to participate of the colours of the Eagle and the Erne: and these differences sufficiently point out this bird as a distinct species. It is probably the most remarkable bird of prey on the Malabar coast, since the natives make an idol of it, to which they pay adoration *; but the beauty of its

* The Malabar Eagle is as beautiful as it is rare. Its head, neck, and the whole of its breast, are covered with exceedingly white feathers, longer than broad, the shaft and edge of which are of a fine jet black. The rest of the body is of shining chesnut, lighter beneath the wings than above; the six first wing-feathers are black at the end; the cere bluish; the tip of the bill yellow, verging on green; the feet yellow; the nails black. It has a piercing look, and is of the bulk of the falcon. It is a sort of divinity worshipped by the people of Malabar. It occurs also in the kingdom of Visapour, and in the territories of the Great Mogul. —SALERNE.

plumage

plumage, rather than its bulk or ſtrength, merits this honour. It is undoubtedly the moſt elegant of the rapacious tribe.

[A] This bird is the Pondicherry Eagle of Latham, and the *Falco Ponticerianus* of Linnæus. " The cere ſky-coloured ; the " body cheſnut ; the head, neck, and breaſt, white, variegated " with duſky lines ; the ſix primary wing-feathers black from " their middle to the tips."

II.

The Bird of South America, deſcribed by Marcgrave under the name *Urutaurana,* which it receives from the Indians in Braſil, and mentioned by Fernandes by the name of *Yſquauthli,* by which it is called in Mexico. It is what our French travellers have termed the *Oronooco Eagle* *, a name which has been adopted by the Engliſh. It is ſomewhat ſmaller than the Common Eagle, and reſembles the Spotted or Rough-footed Eagle by the variety of its plumage. But it has ſeveral ſpecific characters : the tips of its

* The Antilles are often viſited from the continent by a ſort of large bird, which muſt be ranked at the head of the birds of prey in America. The natives of Tobago have called it the *Oronooco Eagle,* becauſe it is of the bulk and the figure of an Eagle ; and becauſe they hold that this bird, which appears only occaſionally in the iſland, frequents the banks of the great river Oronooco. All its plumage is light grey, except the tips of the wings and of the tail, which are edged with yellow. Its eyes are lively and piercing ; its wings very long ; its flight rapid and ſpeedy, conſidering the weight of its body. It ſubſiſts on other birds, on which it darts with fury, tears them in pieces and ſwallows them. It attacks the *arras* and paroquets. It has been obſerved, that it never attacks the bird when on the ground, or ſitting on a branch, but waits till it riſes, and ſeizes it on the wing." Du Tertre, *Hiſt. Nat. des Antilles.*

wings and tail are edged with a whitifh yellow;
two black feathers about two inches long, and two
other fmaller ones, are placed on the crown of
the head, and can be raifed or depreffed at pleafure; the legs are clothed to the feet with white
and black feathers, imbricated like fcales; the
iris is of a bright yellow; the cere and the feet
are alfo yellow, as in the Eagles; but the bill is
of a darker, and the nails of a lighter fhade :—
Thefe differences are fufficient to diftinguifh this
bird from thofe that have been mentioned in
the preceding articles; but to the fame fpecies
we muft, I imagine, refer what Garcilaffo calls
the *Eagle of Peru*, and which, he fays, is fmaller
than the Eagles in Spain.

Of the fame fpecies, or at leaft of a proximate
one, is alfo the bird found on the weft coaft of
Africa, of which Edwards gives an elegant coloured figure, with an excellent defcription, under the name of *Crowned Eagle* *.

The

* This bird, fays Edwards, is about a third fmaller than the largeft
European Eagles, and appears bolder and more intrepid than the
others. The bill with the cere, in which the noftrils are placed, is
of a dull brown; it is cleft as far as the eyes, and the edges of the
mandibles are yellowifh at the infertion; the iris is reddifh orange;
the fore-part of the head, the orbits, and the throat, are covered
with white feathers, fprinkled with fmall black fpecks; the hind
part of the neck and of the head, the back and the wings, are of a
deep brown, verging on black; but the outer edges of the feathers
are light brown. The quill-feathers are of a deeper colour than
the others in the wings; the fides of the wings near the top, and
the ends of fome of the coverts of the wings, are white; the tail is
of a deep grey, interfected with black bars, and the under part
appears

The diftance between Brafil and Africa, which fcarcely exceeds four hundred leagues, is not too great a journey to be performed by a bird of an aërial flight; and therefore it is poffible that it may be found on both coafts. The characters are fufficient to decide the identity of the fpecies; both have a fort of crefts which they can deprefs at pleafure, and both are nearly of the fame fize; in both the plumage is variegated, and fimilarly marked with fpots; the iris is of a bright orange; the bill, blackifh; the legs covered to the feet with feathers, and marked with black and white; the toes yellow, and the nails brown or black. In fhort, the fole difference confifts in the difpofition of the colours, and in the fhades of the plumage, which bears no comparifon to the points of conformity. I fhall not hefitate therefore to confider the birds of the coafts of Africa as of the fame fpecies with that

appears of a dull afh grey; the breaft is of a reddifh brown, with large tranfverfe fpots on the fides; the belly is white, and alfo the under part of the tail, which is marked with black fpots; the thighs and legs are covered to the nails with white feathers prettily marked with round black fpots; the nails are black, and very ftrong; the toes are covered with fcales of a vivid yellow; it erects the feathers on its head like a crown, whence it is named. I drew this bird alive at London in 1752; its owner affured me that it came from the coaft of Africa; and I am the more willing to believe it, as I have two others of the fame fpecies belonging to another perfon, and which were brought from Guinea. Barbot mentions this bird by the name of the *Crowned Eagle*, in his Defcription of Guinea: he has given a bad figure of it; but ftill we may perceive that the feathers rife on the head in a way very little different from that reprefented in mine. EDWARDS.

of

of Brafil; and that the Crowned Eagle of Brafil, the Oronooco Eagle, the Peruvian Eagle, the
Crowned Eagle of Guinea, are all the fame individual, and have the neareft refemblance to
the Spotted or Rough-footed Eagle of Europe.

[A] Linnæus ranges this bird with the Vulture, Gmelin with the
Falcon, and both apply the epithet *Harpyia*; Latham calls it the
Crefted-vulture. It is faid to cleave a man's fkull with one ftroke
of its bill, and to be as large as a ram. There is a variety of this
in New Grenada, which has a black creft, a white belly, thighs
fpotted white, the tail long, and variegated with white and black.
When young it can be tamed.

III.

The Bird of Brazil, mentioned by Marcgrave
by the name of *Urubitinga*, which is probably a
different fpecies from the preceding, fince it receives another name in the fame country. In
fact, it differs, firft, by its fize, being an half
fmaller; fecond, by its colour, being of a
blackifh brown, inftead of a fine grey; third,
by its want of erect feathers on the head;
fourth, becaufe the under part of its legs and
feet are naked, as in the Erne, while the preceding is, like the Eagle, feathered to the talons.

IV.

The Bird which we fhall call the *Little American Eagle*, which has not been defcribed by
any naturalift, and which is found at Cayenne
and other parts of South America, is fcarcely
fixteen or eighteen inches long; and is diftinguifhed at the firft glance by a purplifh red fpot
under

under its neck and throat. It is fo fmall that
we fhould be inclined to clafs it with the Sparrow-hawks or Falcons ; but the fhape of its bill,
which is ftraight at its infertion, and begins its
curve at fome diftance from the origin, has determined us to refer it to the Eagles.

[A] This is the *Red throated Falcon* of Latham, the *Falco Aqui-*
linus of Gmelin, and the *Falco Formofus* of Linnæus. Its fpecific
character is, " That its cere, orbits, and feet, are yellowifh ; its
" throat purple ; its body of a cærulean red ; its abdomen flefh-
" coloured."

V.

The Bird of the Antilles, called *the Fifher* by
Father Tertre, and which is probably the fame
with that mentioned by Catefby by the name
of the *Fifhing-hawk of Carolina*. It is, fays
he, of the fize of a Gos-hawk, but with a
longer body ; its wings, when clofed, ftretch a
little beyond the extremity of the tail, and
when expanded meafure more than five feet.
Its iris is yellow ; its cere blue ; its bill black ;
its feet of a pale blue ; its nails black, and almoft all of the fame length ; the upper part of
the body, of the wings, and of the tail, is dark
brown ; all the under parts are white ; the feathers on the legs are white, fhort, and applied
clofe to the fkin. " The Fifher," fays Father
Tertre, " is exactly like the *Mansfeni*, except that
" its ventral feathers are white, and thofe on
" the crown of the head black ; its claws are
" fomewhat fmaller. This Fifher is a real pi-
" rate ; it molefts not the land-animals, or the
" birds

" birds of the air, but directs its attacks upon
" the fifh alone, which it defcries from the top
" of a branch, or the point of a cliff, and ob-
" ferving them at the furface of the water, it
" inftantly darts upon them, feizes them with
" its talons, and retires to devour its prey on a
" rock. Though it does not wage war againft
" the birds, it is purfued by them, and teafed
" and pecked by them, till it is obliged to fhift
" its place. The Indian children breed them
" when young, and employ them to fifh for
" pleafure merely, for they never give up their
" feizure." This defcription of Father Tertre
is neither fo particular nor fo full as to warrant
us to affert that the bird is the fame with that
mentioned by Catefby ; we fhall therefore ftate
it only as a conjecture. But the American bird
defcribed by Catefby, refembles fo nearly the
European Ofprey, that we are inclined to believe
that it is the fame fpecies, or at leaft only a variety of it. Its colour is nearly the fame ; and
fo are alfo its fize, figure, and habits.

[A] This is the *Carolina Ofprey* of Latham, and is a variety of
the *Falco Haliaëtus* of Linnæus. " Its tail is of a dufky fhade ;
" its crown black, or darkifh, variegated with white ; the belly
" white."

VI.

The Bird of the Antilles, called by our travellers *Mansfeni*, and which they have reckoned a fpecies of the Rough-footed Eagle (*Nifus*).
The *Mansfeni*, fays Father Tertre, is a ftrong
bird of prey, which in its form and plumage
bears

bears fo great a refemblance to the Eagle, that its diminutive fize is the only mark of difcrimination, for it is fcarcely bigger than the Falcon, but its claws are twice as large, and ftronger. Though thus well-armed, however, it generally attacks only the defencelefs birds; as the thrufhes, and the fea-larks; or if more adventurous, the ring-doves and turtles : it feeds alfo on ferpents and fmall lizards. It perches commonly on the moft lofty trees. Its feathers are fo ftrong and fo compacted, that unlefs we fire oppofite their pofition, the fhot will not penetrate. Its flefh is rather black, but yet of a pleafant flavour.

The VULTURES*.

THE Eagles have been placed at the head of the birds of prey, not becaufe they are larger or ftronger than the Vultures, but becaufe they are more generous, that is, not fo meanly cruel; their difpofitions are bolder, their conduct more intrepid, and their courage nobler. They are ftimulated to their attacks, as much from the glory of conqueft as the appetite for plunder: the Vultures, on the contrary, are incited by a low gormandizing inftinct; and they feldom attack living animals when they can fatiate their voracity on the carcafes of the dead ones. The Eagle makes a clofe fight with his enemies or his victims; he purfues his prey alone and unaided, and fingly ravifhes the plunder, contends with, and fecures his prey. The Vultures, on the contrary, when they expect

* [A] The following is the Generic Character of the Vultures, as given by the beft fyftematic Writers:

The BILL ftraight, blunt at the tip, and the bafe covered with fkin.
The HEAD moftly deftitute of feathers, and naked fkin on the forepart.
The TONGUE flefhy, generally cloven.
The NECK retractile.
The FEET ftrong, and the nails moderately incurvated.

the

THE VULTURE

the flighteft refiftance, combine in flocks, like
bafe affaffins, and are rather robbers than war-
riors, birds of carnage than birds of prey. This
tribe alone collect in numbers to pour upon the
forlorn individual, and tearing the mangled
carcafe to the bones, they difplay the bitternefs
of unprovoked rage. Corruption and infection,
inftead of driving them to a diftance, are to them
powerful attractions. Sparrow-hawks, falcons,
and even the fmalleft birds fhew more courage;
for they feek their prey alone, and almoft all of
them reject putrid flefh, and fpurn a dead carcafe. If we compare birds with the quadrupeds,
the Vulture feems to unite the ftrength and
cruelty of the tiger with the cowardice and
voracity of the jackals, which gather in troops
to devour carrion, and dig up carcafes; while
the Eagle poffeffes, as we have faid, the courage,
the generofity, the magnanimity, and the munificence of the lion.

We muft therefore feparate the Vultures from
the Eagles by this difference of inftinct; and
their external appearance fufficiently marks the
diftinction. Their eyes are raifed, while thofe
of the Eagle are funk in the orbit; the head is
bare, the neck almoft naked, or covered with a
flight down, or fprinkled with a few ftraggling
hairs; while the Eagle is completely clothed with
feathers: the nails of the Eagle are almoft femicircular, fince they feldom reft upon the ground;
while thofe of the Vulture are fhorter and lefs

9   curved;

curved : their posture is more inclined than that
of the Eagle, which is boldly erect, and almost
perpendicular upon its feet ; while the Vulture,
whose situation is half horizontal, seems to be-
tray the baseness of its character by the inclined
position of its body.   The Vultures can even
be distinguished at a distance ; because they are
the only birds of prey that fly in flocks, that is,
more than two or three together : besides their
flight is slow and laborious ; it is painful for
them to rise from the ground ; and they are
obliged to make three or four attempts before they
can succeed *.

We have included in the genus of Eagles
three species, viz. the Golden Eagle, the Ring-
tail Eagle, and the Rough-footed Eagle ; and
we have added those birds which bear the greatest
resemblance to them ; such as the Erne, the Osprey, the Sea-Eagle, and the White John,
and the foreign birds related to these : viz. 1.
The Beautiful Bird of Malabar.   2. The Bird

---

* Ray, and Salerne, (who generally copies him word for word,)
add another discriminating difference between the Vultures and the
Eagles, viz. the shape of the bill, which is not curved immediately at its origin, but continues straight for two inches before it
bends. But I must observe, that this character is not precise ; for in
the Eagles also, the bill does not immediately begin its arch ; the
real difference is, that the straight part is longer in the Vultures
than in the Eagles.   Other naturalists assign the prominence of the
craw as the character of the Vultures ; but this is equivocal, and
does not belong to all the species.   The Fulvous Vulture, which
is one of the principal, so far from having a projection, is remarkable for a concavity under its neck,

of

of Brazil, Oronooco, Peru, and Guinea, called by
the Indians of Brazil *Urutauana*. 3. The Bird
called in the fame country *Urubitinga*. 4. That
which we have called the *Little American Eagle*.
5. The Bird Fisher of the Antilles. 6. The
Mansfeni, which appears to be a kind of the
Rough-footed Eagle:—thefe conftitute in all
thirteen fpecies, of which the *Little Eagle of
America* has been mentioned by no naturalift.
We proceed to make in the fame manner the
enumeration and reduction of the fpecies of the
Vultures, and we fhall firft treat of a bird which
has been ranked among the Eagles by Ariftotle,
and after him by moft authors, though it is
really a Vulture.

The ALPINE VULTURE.

Le Percnoptere, Buff.
Vultur Percnopterus, Linn.
Falco Montanus Ægyptiacus, Haffelq.
Vultur Aquilina, Alb.

THIS bird is by no means an Eagle, and is certainly a Vulture; or if we could follow the opinion of the ancients, it forms the laft fhade between thefe two kinds of birds, bearing a much clofer refemblance to the former than to the latter. Ariftotle *, who ranges it among the Eagles, confeffes himfelf that it is rather of the Vulture race, having, he fays, all the bad qualities of the Eagles without any of their virtues; fuffering itfelf to be purfued and haraffed by the crows; indolent in the chace, and tardy in its motions; always crying and complaining; always famifhed and fearching for carrion. Its wings are alfo fhorter, and its tail longer than the Eagles; its head is of a fine blue, the neck white and naked, or covered merely with a

* Ariftotle makes it his fourth fpecies of Eagle under the name of Περκνοπlερος, and afterwards applies the epithet of Υπαεlος, which Theodore Gaza properly tranflates *Subaquila*. But others, and particularly Aldrovandus, have conjectured, that inftead of Υπαεlος we ought to read Γυπαεlος, or *Vulturina aquila*. The fact is, that both thefe appellations fuit this bird.

hoary

hoary down, with a collar of fmall white hard
feathers below the neck like a ruff; the iris is of
a reddifh yellow; the bill and the cere black,
the hook of the bill whitifh; the lower part of
the feet and legs naked, and of a leaden colour;
the nails are black, fhorter, and ftraighter than
thofe of the eagle. It is remarkable for a brown
fpot fhaped like a heart, and edged with a
ftraight white line, and fituated on the breaft
under the ruff. In general, this bird is of an
ugly and ill proportioned figure; it has even a
difgufting appearance from the continual flux of
rheum from its noftrils, and the flow of faliva
from two other holes in the bill; its craw is prominent; and when it is upon the ground, it
keeps its wings always extended *. In fhort,
it refembles the Eagle only by its fize; for it is
larger than the Ring-tail Eagle, and approaches
the Golden-eagle in the thicknefs of its body,
though the expanfion of its wings is lefs. This
fpecies feems to be more rare than thofe of the
other Vultures; it is found however in the
Pyrenees, the Alps, and the mountains of
Greece. [A]

* This habit of holding the wings extended is not peculiar to
this fpecies, but belongs to moft of the Vultures, and many other
birds of prey.

[A] Buffon feems to range with the Alpine Vulture another,
which the fyftematic writers reckon a different fpecies. It is the
Vulturine Eagle of Albin, the *Vultur Barbatus* of Linnæus, the Alpine Vulture of Briffon, the *Percnopterus Gypaëtos* of Ray, which
Buffon mentions in the preceding note. " It is whitifh flame-
" coloured, the body dufky, with a black ftripe above and below
" the eyes."

The FULVOUS VULTURE.

Le Griffon, Buff.
Vultur Fulvus, Linn.
Vultur Ruber, Rzac.

THIS bird is ftill larger than the Alpine Vulture; its wings extend eight feet; its body
is thicker and longer than that of the Golden-
eagle, the legs being more than a foot in length,
and the neck feven inches. It has, like the
Alpine Vulture, a ring of white feathers at the
origin of the neck; the head is covered with
fimilar feathers, which are collected into a tuft,
under which can be perceived the perforations
of the ears; the neck is entirely deftitute of
plumage; the eyes are level with the head, with
large eye-lids, which are moveable, and furnifhed
with lafhes; the iris is of a beautiful orange
colour; the bill long and hooked, black at its
origin and termination, and bluifh in the middle.
The bird is alfo diftinguifhed by a re-entrant
craw, or a large cavity above the ftomach, which
cavity is covered with hairs, pointing to its centre,
and occupying the place of the craw; it is neither
prominent nor pendulous, as the Alpine Vulture.
The fkin, which appears naked on the neck,
round the eyes, ears, &c. is of a brown grey,
and

and bluifh; the largeft feathers of the wing are
two feet long, and the quill is an inch in circumference; the nails are blackifh, but not fo
large or fo crooked as thofe of the eagles.

I believe, as the Members of the Academy
of Sciences have faid, that the Fulvous Vulture
is really the Great Vulture of Ariftotle; but,
as they give no reafons in fupport of their opinion, and as Ariftotle feems to form only two
fpecies, or rather genera of Vultures, the Little
one being whiter than the Great, which differs
alfo in its form *; it would appear that this
genus of the Great Vulture includes more than
one fpecies. For there is only the Alpine Vulture which he particularly mentions, and as he
does not defcribe any of the other Great Vultures,
we may reafonably doubt if the Fulvous Vulture
was the fame with his Great Vulture. The Common Vulture, which is as large, and perhaps
more common than the Fulvous Vulture, might
be equally taken for this Great Vulture; fo that
we may infer that the Members of the Academy
of Sciences were rafh in affirming as certain, a
thing fo equivocal and fo doubtful, without even
mentioning the reafon or ground of their affertion; which may be perhaps true, but which
muft be proved by reflections and comparifons
which they have not made. I fhall endeavour

* There are two forts of Vultures, the one fmaller and whiter,
the other larger and more variegated. ARIST. *Hift. Anim.*

to

to perform this talk; and shall here state the reasons which have convinced me that the Fulvous Vulture is really the Great Vulture of the ancients.

I am then of opinion, that the species of Fulvous Vulture consists of two varieties; the first called by naturalists the *Tawny Vulture* *, and the second the *Golden Vulture* †. The difference between these two birds, of which the first is the Fulvous Vulture, is not so considerable as to constitute two distinct species, for both are of the same size, and nearly of the same colour; in both, the tail is comparatively short, and the wings very long ‡, and by this common character they are distinguished from the other Vultures. This close resemblance § has struck some naturalists even before me, and has induced them to reckon these kindred species. I am even inclined to believe that the bird mentioned by Belon, under the name of *Black Vulture*, is still of the same species with the Golden and Fulvous Vulture; for it is of the same bulk, and its

* *Le Vautour Fauve.* Briss.

† *Vultur Aureus*, Alberti Magni, Gesnerii, Raii, Willoughbei, Klein. *Vultur Bæticus, sive Castaneus*, Alb. *Le Vautour Doré*, Briss.

‡ Brisson states the tail of his Golden Vulture as of two feet three inches long, and the largest wing-feather three feet; which would make me suspect that it is a different bird from the Golden Vulture of the ancients, whose tail was shorter compared with its wings.

§ Ray and Willoughby.

back

back and wings have the fame colour as in the
Golden Vulture. But if we unite thefe three
varieties into one fpecies, the Fulvous Vulture
would be the leaft unfrequent of all the Great
Vultures, and confequently that which Ariftotle
would principally mention. And what adds
probability to the prefumption is, that, according to Belon, this Great Vulture is found in
Egypt, Arabia, and the iflands of the Archipelago,
and therefore common in Greece. At any rate
I am confident that we may reduce the Great
Vultures which appear in Europe into four fpecies :—the Alpine, the Fulvous, the Vulture
properly fo called, of which we fhall treat in
the following article, and the Crefted Vulture ;
which differ fufficiently from each other to conftitute feparate and diftinct fpecies.

The Academicians, who diffected two female
Fulvous Vultures, have well obferved, that the
bill is longer, and lefs incurvated than in the
Eagles ; and that it is black only at the origin
and the tip, the middle being of a bluifh grey ;
that the fuperior mandible is marked within
with a groove on each fide ; that thefe receive
the cutting edges of the inferior mandible when
the bill is clofed ; that towards the point of the
beak there is a fmall round protuberance, on the
fides of which are two little perforations through
which the faliva is difcharged ; that at the bafe
of the beak are placed the two noftrils, each

fix lines long and two broad, meafuring downwards, which gives an ample fpace for the
external organs of fmell; that the tongue is
hard and cartilaginous, fcooped near the tip,
and the edges raifed; that thefe raifed edges are
ftill harder than the reft of the tongue, and
form a kind of faw, the teeth of which are
pointed towards the gullet; that the œfophagus
dilates below, and forms a large fac; that this
fac differs from the crop of fowls only becaufe
it is interfperfed with the ramifications of a great
number of veffels which are very diftinct, the
membrane being exceedingly white and tranfparent; that the gizzard is neither fo hard nor
fo thick as in the gallinaceous tribe, and that
the flefhy part is not fo red as in the gizzards of
other birds, but white, like the ventricles; that
the inteftines and the *cæcum* are fmall as in other
rapacious birds; and that the *ovarium* is of the
ordinary fhape and fize, and the *oviductus* fomewhat ferpentine, as in the poultry, and does
not form a ftraight regular canal as in moft
other birds.

If we compare thefe obfervations on the interior ftructure of Vultures with thofe which the
fame anatomifts of the Academy made on Eagles,
we fhall cafily perceive, that though the Vultures
feed upon flefh, as do the Eagles, they have
not the fame conformation in the organs of digeftion; and that, in this refpect, they approach
much

much nearer to the poultry and other birds that live upon grain; fince they have a craw and a ftomach which, from the thicknefs of its lower part, may be regarded as almoft a gizzard; fo that the Vultures feem deftined by their ftructure, not only to be carnivorous, but gra-nivorous, and even omnivorous.

The CINEREOUS VULTURE.

Le Vautour, ou *Grand Vautour,* Buff.
Vultur Cinereus, Linn.
In Italian, *Avoltorio.*
In Spanifh, *Buyetre.*
In German, *Geyr.*
In Polifh, *Sep.*
In Arabic, *Racham.*

THIS bird is thicker and larger than the Common Eagle, but rather fmaller than the Fulvous Vulture, from which it is not difficult to diftinguifh it :—1. Its neck is covered with a longer and thicker down, of the fame colour with that of the feathers on the back ; 2. It has a fort of white collar, which rifes on both fides of the head, and extends in two branches to the bottom of the neck, bordering on each fide a pretty broad black fpace, under which is a narrow white ring ; 3. Its feet are covered with brown feathers, while, in the Fulvous Vulture, they are yellowifh or whitifh ; and, 4. The toes are yellow ; whereas in the Fulvous Vulture they are brown, or of an afh colour.

The HARE VULTURE.

Le Vautour à Aigrettes, Buff.
Vultur Criftatus, Linn.
Vultur Leporarius, Ray, Will. and Klein.

THIS Vulture, though fmaller than the three firft, ftill deferves to be ranked among the Great Vultures. We cannot defcribe it better than in the words of Gefner, who is the only naturalift that has feen many of thefe birds. The Vulture, fays he, which the Germans call *Hafengeier*, *(Hare Vulture,)* has a black bill, hooked at the point, ugly eyes, a large and ftrong body, broad wings, a long and ftraight tail; a blackifh rufty plumage, and yellow feet. When at reft, whether on the ground or perched, it erects the feathers of the head, which then refemble two horns, but which are not perceived when it flies. The expanfion of its wings is near fix feet; it walks well, advancing fifteen inches at each ftep; it purfues birds of every kind, and preys upon them; it alfo catches hares, rabbits, young foxes, and fmall fawns, nor does it fpare even the fifh. Its ferocity is fuch, that it cannot be tamed. Sometimes it feizes its prey in its flight; at other times, it darts upon

I 3

its

its victims from the top of a tree or elevated cliff; but always upon the wing. It makes much noise in its flight. It breeds in the thick and desert forests on the tallest trees. It eats flesh, the entrails of living animals, and even carrion; and though extremely voracious, it can bear the want of food for fourteen days. Two of these birds were caught in Alsace in the month of January 1513; and in the following year, more were found in a nest built on a thick lofty oak, at some distance from the city of Misen.

All the Great Vultures, including the Alpine, the Fulvous, the Cinereous, and the Hare Vulture, have but few young, and breed only once a-year. Aristotle * says, that they generally lay only one or two eggs. They build their nests in places so lofty and inaccessible, that they are seldom discovered; they must be sought for only on the giddy heights of desert mountains †. The Vultures prefer the gloomy haunts during the

---

* " It breeds on inaccessible rocks, nor is this bird an inhabit-
" ant of many places, for it has only one or two young." ARIST,
Hist. Anim.

† In general, Eagles and Vultures which inhabit islands and tracts near the sea, do not build upon trees, but establish their abode among frightful precipices, so that they cannot be seen except from on board of a ship. BELON.—Dapper gives the same relation, and adds, that a strong stake is driven firmly into the earth that covers the summit of the rock, and that a man is let down by a long rope fixed to it, and after putting into a basket the eggs or young which he finds, he is drawn up by his companions. [It is well known that this is practised in the western islands of Scotland and in the Orknies.]

whole

whole of the fine feafon; but when fnow and
ice begin to cover the fummits of the mountains,
they defcend into the plains, and feek more hof-
pitable abodes. Vultures feem to dread more
than Eagles the influence of cold; they are lefs
common in the north, and it would feem that
they have not penetrated Sweden, or the more
diftant boreal regions; for Linnæus, in the enumeration which he has given of the Swedifh
birds, makes no mention of the Vultures: however, in the following article, we fhall defcribe
a Vulture which we have received from Norway.
But they delight in warm climates; and in Egypt *,
Arabia, the iflands of the Archipelago, and other
parts of Africa and Afia, they are numerous.
In thofe countries the natives make great ufe of
Vultures' fkins; the leather is almoft as thick as
that of the kid, and covered with a fine, clofe,
warm down, and they manufacture it into excellent furs †.

It

* When I was in Egypt and in the plains of Arabia Deferta, I
obferved that the Vultures were frequent and large. BELON.

† The peafants of Crete, and others who inhabit the mountains in different countries in Egypt and in Arabia Deferta, con-
trive to catch the Vultures feveral ways; they flay them, and fell
their fkins to the furriers. . . . Their fkin is about as thick as that
of a kid . . . The furriers can feparate the large feathers, and leave
the down, which is under thefe, and thus currying the fkin, they
manufacture valuable *peliffes*; but in France it is generally ufed for
making ftomachers. . . . A perfon who happens to be at Cairo,
and view the merchandizes that are expofed to fale, will find delicate filky fur veftments made of Vultures fkins, both black and
white,

It appears to me that the Black Vulture, which
Belon fays is common in Egypt, is one of the
fame fpecies with the Cinereous Vulture, and
that we ought not to feparate them, as fome
naturalifts have done ; fince Belon, who alone
has mentioned them, does not diftinguifh them,
and fpeaks of the Cinereous and the Black as
compofing the fpecies of the Great Vulture.   In
fhort, it is probable that there are really black
ones, fuch as figured N° 425, and others that
are Cinereous, but which we have not feen.
The Black Vulture and the Black Eagle are both
of the common fpecies of the Vulture or the
Eagle.   Ariftotle properly remarked, that the
genus of the Great Vulture was various ; for
without including the Alpine, which removes
from the Vultures, and ranges with the Eagles,
it is really compofed of three fpecies ; the Fulvous,

white. . . . . There are many Vultures in the ifland of Cyprus;
thefe birds are of the bulk of a fwan, very like the Eagle, the
wings and back being covered with the fame fort of feathers ;
their neck is full of down, foft as the fineft fur, and all the fkin
fo covered with it, that the iflanders wear it upon their breaft, and
oppofite to their ftomach, to affift digeftion. Thefe birds have a
tuft of feathers under their neck ; their legs are thick and ftrong.
. . . . They feed only on carrion, and gorge themfelves fo much,
that one meal is fufficient for fifteen days. . . And when thus
gorged, they cannot eafily rife from the ground ; they are then
readily fhot. Sometimes even they are fo inactive, that they can
be caught by dogs, or killed with fticks or ftones. *Defcription de
l'Archipel, par* DAPPER.

the

the Cinereous, and the Hare Vultures. The Little, or Aſh-coloured Vulture, on the contrary, forms a ſingle ſpecies only in Europe; and thus the philoſopher had ſtill reaſon to ſay, that the genus of the Great Vulture was more varied.

The ASH-COLOURED VULTURE.

Le Petit Vautour, Buff.

WE have now to confider the Small Vultures, which differ from thofe of which we have already treated, not only by their diminutive fize, but by other peculiar characters. Ariftotle reckons only one fpecies, but our nomenclators make three; the Brown Vulture, the Egyptian Vulture, and the White-headed Vulture. The laft *, which is the fmalleft, appears really to be of a different fpecies from that of the two firft; for its legs and feet are naked, while in the others they are feathered. This White-headed Vulture is probably the Little White Vulture of the Ancients, which was common in Arabia, Egypt, Greece, Germany, and even as far as Norway, whence it was fent to us. We may remark, that the head, and the under part of the neck, are naked, and of a reddifh colour; and that the bird is entirely white, excepting the large feathers of the wings, which are black †.—Thefe characters are full fufficient to difcriminate it.

* *Vultur Leucocephalus*, LINN. and BRISS. and the *White Vulture* of RAY. " Its body is footy, with rufous fpots; its head, neck, " and bottom of the tail, white." It is two feet fix inches long.

† Schwenckfeld fays, that in Silefia it is called *Grimmer*; that its tongue is pretty broad; that its ftomach is thick and wrinkled; and its gall-bladder large.

Of

Of the other fpecies of the Afh-coloured Vulture, I am inclined to reject, or rather to feparate, the fecond, which, from Belon's defcription, is not a Vulture, but a bird of another
genus, which he calls the *Egyptian Sacre.*
There remains therefore only the Brown Vulture ; with regard to which I fhall obferve,
that I cannot perceive the reafons which led
Briffon to refer it to the *Aquila Heteropos*
of Gefner.   On the contrary, it appears to me,
that inftead of reckoning the Heteropede Eagle
a Vulture, we ought to erafe it from the catalogue of birds ; for its exiftence was never
proved ; no naturalift has feen it ; and Gefner,
who is the only one that mentions it, and
whom all the others have copied, had only a
drawing of it, which he caufed to be engraved,
and from the figure, referred it to the genus of
Eagles, and not to that of Vultures ; and the
epithet of *Heteropede* alludes to the circumftance
that one of the legs was blue, and the other
whitifh brown.   But a bird, figured by an unknown perfon, and named afterwards from an
inaccurate drawing, and which the difference of
colour of the legs is alone fufficient to render
fufpicious ; a bird which has never been feen by
thofe who mention it ; can we confider fuch as
an Eagle or a Vulture ? or has it any real exiftence ? It appears then, that to refer it to the
Brown Vulture is mere hypothefis,

FOREIGN BIRDS,
WHICH RESEMBLE THE VULTURES.

I.

THE bird fent from Africa, and the ifle of Malta, under the name of *Brown Vulture*, mentioned in the preceding article, which is a particular fpecies or variety of the Vulture tribe, and which is not found in Europe, muft be confidered as a native of the climate of Africa, efpecially of the countries bordering on the Mediterranean.

II.

The bird called by Belon the *Egyptian Sacre**, and which Dr. Shaw mentions by the name of *Achbobba*. This bird appears in numerous flocks on the barren and fandy tracts near the pyramids of Egypt. It is almoft always on the ground, and feeds like the Vultures upon every kind of flefh and carrion. " It is," fays Belon, " a dirty and a vulgar bird ; and whoever will " picture in his imagination a bird with the " bulk of the kite, with a bill intermedi- " ate between the raven and a bird of prey, " hooked at the point, and refembling the ra-

---

* This is a variety of the Alpine Vulture *(Vultur Percnopterus,* LINN.) :—" It is of a rufous afh-colour, with dufky fpots ; its feet " naked."

" ven

" ven in the legs and feet and in the manner
" of walking, will have an idea of this bird,
" which is common in Egypt, and occurs fel-
" dom in any other part of the world ; though
" there are fome in Syria ; and I myfelf have
" feen feveral in Caramania." This bird varies
in its colours. Belon conceives that it is the
Hierax, or the *Egyptian Hawk* of Herodotus, which, like the Ibis, was held in venera-
tion by the ancient Egyptians, becaufe both of
them deftroy and eat the ferpents, and other noxious and difgufting reptiles which infeft Egypt.
" Near Cairo," (fays Dr. Shaw, vol. ii. p. 449.)
" there are feveral flocks of the *Ach bobba**,
" the *Percnopterus*, or *Oripelargus* †, which, like
" the ravens about London, feed upon the car-
" rion and naftinefs, that is thrown without the
" city. The fame bird likewife might be the
" *Egyptian Hawk*, which Strabo defcribes (con-
" trary to the ufual qualities of birds of that
" clafs) to be of no great fiercenefs." Paul
Lucas alfo fpeaks of this bird. " There are
" in Egypt," fays he, " thofe Hawks which
" were honoured, like the Ibis, with religious

* *Ach bobba*, in the Turkifh language, fignifies *White father* ;
a name given it partly out of the reverence they have for it, partly
from the colour of its plumage : though, in the other refpect, it differs little from the *Stork*, being black in feveral places. It is as
big as a large capon, and exactly like the figure which GESNER,
lib. iii. *De Avib.* p. 176, hath given us of it.

† Vid. GESN. ut fupra. ARIST. *Hift. Anim.* lib. ix. cap. 32.
PLIN. lib. x. cap. 3.

" adora-

" adoration. It is a bird of prey of the bulk
" of a raven, the head resembling that of the
" Vulture, and the feathers those of the Fal-
" con.   The priests of this country conceal
" great mysteries under the symbol of this bird.
" They carve the figure on their obelisks and
" the walls of their temples, to represent the
" sun.   The vivacity of its eyes, which it di-
" rects constantly to that great luminary, the
" rapidity of its flight, its longevity, &c. seem
" proper to mark the nature of the star of the
" day," &c. But this bird, which we see is but
imperfectly described, is perhaps the same with
the Carrion Vulture, of which we shall treat in
Art. IV.

III.

Vultur Papa, Linn.
Vultur Monachus, Klein.
Rex Vulturum, Briss.
Cozcacoauhtli, Ray.
Queen of the Auræ, Will.
King of the Vultures, Edw. Alb. & Lath.

The bird of South America which the European settlers have called the *King of the Vultures*,
and which is indeed the most beautiful of the
genus.   Brisson describes it fully and accurately
from a specimen in the Royal cabinet ; and Edwards, who saw several of the birds in London,
has given an excellent drawing of it, and a correct
description.   We shall here collect the remarks
of these authors, and those of their predecessors,
and add the observations which we have our-

selves

THE KING OF VULTURES.

felves made relating to the ftructure and nature
of this bird. It is undoubtedly a Vulture ; for
its head and neck are naked, which is the moft
difcriminating character of the genus. But it is
not large, the extreme length of its body not
exceeding two feet two or three inches ; it is
not thicker than the female turkey ; its wings
are fhorter in proportion to the other Vultures ;
its bill is thick and fhort, and begins its curvature only at the point ; in fome the bill is entirely red ; in others only red at the extremity,
and black in the middle ; the cere is of an
orange-colour, broad, and ftretching from each
fide to the crown of the head ; in the cere are
placed the noftrils, of an oblong fhape, and between which the fkin projects like a loofe jagged comb, falling indifferently on either fide, according as the bird moves its head ; the eyes
are furrounded by a red fkin, and the iris has
the colour and luftre of pearls ; the head and
neck are naked, the crown covered with a flefh-
coloured fkin, which is of a lively red behind,
and darker before ; below the hind part of the
head rifes a fmall tuft of down, from which
there extends, on each fide under the throat, a
wrinkled fkin of a brownifh colour, and mixed
with blue and red near its termination ; this
fkin is marked with fmall lines of black down.
The cheeks, or fides of the head, are covered
with a black down ; and between the bill and
the eyes, behind the infertion of the mandibles,
there

there is a brown purple spot; on the upper part of the arch of the neck there is on each side a small longitudinal line of black down, and the space included between these two lines is of a dirty yellow; the sides of the arch of the neck are of a red colour, which, as it descends, passes by insensible shades into yellow; under the naked part of the neck is a collar or ruff composed of pretty long soft feathers of a deep ash-colour; this collar, which entirely encircles the neck, and descends upon the breast, is so broad that, when the bird contracts itself, it can conceal the neck and part of the head like a cowl; and this is the reason why some naturalists have given it the name of *Monk*. The feathers on the breast, the belly, the thighs, the legs, and the under surface of the tail, are white, slightly tinged with yellow; those of the rump, and the upper surface of the tail, vary, being black in some individuals, and in others white: the other feathers of the tail are always black, and so are the great feathers of the wings, which are commonly edged with grey. The colour of the feet and nails is not the same in all these birds; in some the feet are of a dull white, or yellowish, and the nails blackish; in others, the feet and nails are reddish; the nails are very short, and but slightly curved.

This bird is a native of South America, and not of the East Indies, as some authors have asserted.

afferted *. The fpecimen in the king's cabinet
was fent from Cayenne. Navarette, fpeaking
of this bird, fays, " I faw at Acapulco the
" King of the *Zopilotes*, or *Vultures*; it is one
" of the moft beautiful of birds," &c. Perry, who
dealt in foreign animals at London, informed
Mr. Edwards, that this bird comes only from
America. Hernandes, in his Hiftory of New
Spain, defcribes it in a manner that cannot be
mifunderftood. Fernandes, Nieremberg, and
Laet †, who have all copied the defcription of
Hernandez, agree with him in faying, that this
bird is common in Mexico and New Spain; and
as, in the extenfive fearch which I have made
in works of travellers, I have not difcovered

---

* Albin fays, that the one which he figured was brought in a
Dutch vèffel from the Eaft Indies. And Edwards tells us, that the
perfons who fhewed thefe birds in London affured him, that they
came from the Eaft Indies; notwithftanding, he fuppofed them to
be American.

† " In New Spain there are an incredible number and variety
of beautiful birds, among which are the *Cofquauhtli*, or *Aura*, as the
Mexicans call it. It is of the bignefs of an Egyptian hen, and its
feathers are black upon every part of the body, except at the neck
and round the breaft, where they are of a reddifh black; the wings
are black, and mixed with cinereous, purple, and tawny; the nails
are reflected; the bill red at the point; the noftrils open; the eyes
black, the pupils tawny; the eye-lids red; the face blood-coloured,
and filled with many wrinkles, which it contracts and clofes like the
turkies, and where there is alfo a little woolly hair like Negroes;
the tail is like that of the eagle, black above, and cinereous below."
—" There is another bird of the fame kind, which the Mexicans
call *Tzopilotl*." De Laet, *Hift. du Nouv. Monde*.

This fecond bird, called *Tzopilotl*, muft be a Vulture; for the
King of the Vultures is alfo named by the Mexicans *King of the
Tzopilotles*.

the flighteft indication of it among the birds of
Africa and Afia, I think we may conclude, that
it is peculiar to the Southern regions of the
New Continent, and is not found in the Old.
It may be objected, that fince the *Ouroutaran*,
or Eagle of Brazil, frequents, as I admit, both
the African and American fhores, the King of
the Vultures may enjoy the fame extenfive
range.   But this bird is probably unequal to the
journey*; for the Eagles in general fly better
than the Vultures.   It is delicately fenfible of
cold, and therefore could not pafs by the way of
the North.   I am at leaft certain, that this bird
is confined to its natal region, and haunts the
tracts between Brazil and New Spain.

The King Vulture is neither elegant, noble,
nor generous; it attacks only weak victims, and
feeds upon rats, lizards, ferpents, and even the
excrements of animals and men.   Hence it has
a difgufting fmell, and not even the favages can
eat its flefh.

IV.

The bird called *Ouroua*, or *Aura*, by the In-
dians of Cayenne, *Urubu* by thofe of Brazil,
Zopilotl by thofe of Mexico, and to which the
French fettlers in St. Domingo have applied the

---

* Fernandez however fays, that this bird rifes to a great height,
holding its wings much extended; and that its flight is fo vigorous
that it withftands the greateft winds.   One might fuppofe that
Nieremberg alluded to this circumftance when he called it *Regina
Aurarum*; but *Aura* is not derived from the Latin, it is a contraction for *Ouroua*, the Indian name of the Carrion Vulture.

epithet

epithet of *Merchant*, alfo muft be referred
to the genus of Vultures; for it has the fame
inftinctive difpofitions, and, like them, its bill is
hooked, and its head and neck deftitute of plumage. It bears alfo fome refemblance to the
turkey, which has occafioned its receiving from
the Spaniards and Portuguefe the name of *Gal-
linaço* *. It hardly exceeds the fize of the wild
goofe; its head appears fmall, becaufe it, as well
as the neck, is covered only with naked fkin,
with fome ftraggling black hairs; the fkin is
rough, and variegated with blue, white, and
red; the wings, when clofed, extend beyond
the tail, which is alfo of confiderable length;
the bill is of a yellowifh white, and curved only
at the point; the cere extends almoft to the
middle of the bill, and is of a reddifh yellow;
the iris is orange, and the eyelids white; the
feathers are brown or blackifh over the whole
body, and reflect a varying colour of dull green
and purple; the feet are of a livid colour, and
the nails black. This bird has noftrils ftill
longer in proportion than the other Vultures †;

* *Vultur Aura*, Linn. *Strunt-vogel* ||, Kolb. *Turkey-Buzzard*,
Catefby and Clayton. *Carrion-Vulture*, Sloane, Damp. Brown,
Penn. and Lath.

" It is of a dufky grey; the wing-feathers black; and the bill
white."

† I have thought it proper to give this fhort defcription, becaufe
thofe of other authors do not agree precifely with one I have feen.
But as the differences are flight, we may prefume that they were
owing to the peculiarities of the individual, and confequently their
defcriptions may be as good as mine.

|| (i. e. *nafty bird.*)

K 2

it

it is accordingly more cowardly, more filthy,
and more voracious than any of them, feeding
rather upon carrion and filth than upon
game. Its flight, however, is lofty and rapid;
but it has not courage to pursue its prey, and
only grovels among the dead carcasses. If it
sometimes summons resolution to make an as-
fault, it collects in numerous flocks, and sur-
prises the helpless solitary animal while drowned in sleep or disarmed by wounds.

The Carrion Vulture is the same bird with
that which Kolben describes under the name of
the *Eagle of the Cape.* It is therefore found
both on the continent of Africa and that of South
America; and as it is not observed in the countries of the North, it must have traversed the sea
between Brazil and Guinea. Hans Sloane, who
saw many of them in America, says, that they fly
like kites, and are always lean. Hence it is very
possible, from their agility and the rapidity of
their course, that they could perform the distant
journey across the ocean which separates the two
continents. Hernandes informs us, that they feed
upon animal carcasses, and even human excrements; that they assemble on the lofty trees,
whence they descend in flocks to devour carrion;
and he adds, their flesh has an offensive smell,
ranker than that of the raven. Nieremberg also says,
that they fly very high and in numerous flocks; that
they pass the night upon trees or elevated rocks,
which they leave in the morning, and resort near

the

the inhabited fpots; that their fight is very acute, and that they defcry, from a towering height and an immenfe diftance, the carcaffes on which they prey; that they maintain a gloomy filence, and never fcream or fing, and are heard only by a flight murmur, which they feldom utter; that they are very common on the plantations in South America, and that their young are white in their infancy, and become brown or blackifh as they grow old. Marcgrave, in the defcription which he gives of this bird, fays, that its feet are whitifh, its eyes bright, and of a ruby colour; the tongue grooved, and ferrated on the fides. Ximenes affures us, that thefe birds never fly but in large flocks, and are always very lofty in their courfe; that they all alight together upon the fame prey, and, without contention, devour it to the bones, and gorge themfelves to fuch a degree, that they are unable to refume their flight. Thefe are the fame birds that Acofta mentions by the name of *Poullazes,* " which " have," fays he, " a wonderful agility and a pier- " cing eye, and are very ufeful for cleaning cities, " not fuffering the leaft veftige of carrion or " putrid matter to remain; that they fpend the " night upon trees and rocks, and refort to the " towns in the morning, perch upon the top " of the higheft buildings, whence they defcry " and watch for their plunder; their young " have a white plumage, which afterwards " changes with age into black."—"I believe," fays

Defmarchais,

Defmarchais, " that thefe birds called *Gallinaches*
" by the Portuguefe, and *Marchands* by the
" French fettlers in St. Domingo, are a kind of
" turkey *, which, inftead of living upon grain,
" fruits, and herbs, like the others, are accuf-
" tomed to feed upon dead animals and carrion;
" they follow the hunters, efpecially thofe whofe
" object is only to procure the fkins; thefe
" people neglect the carcaffes, which would rot
" on the fpot, and infect the air, but for the
" affiftance of thefe birds, which no fooner per-
" ceive a flayed body, than they call to each other,
" and pour upon it like Vultures, and in an in-
" ftant devour the flefh, and leave the bones as
" clean as if they had been fcraped with a knife.
" The Spaniards, who are fettled upon the large
" iflands, and upon the continent, as well as the
" Portuguefe, who inhabit thofe tracts where
" they traffic in hides, receive great benefit
" from thefe birds, by their devouring the
" dead bodies and preventing infection; and
" therefore they impofe a fine upon thofe who
" deftroy them. This protection has extremely
" multiplied this difgufting kind of turkey. It
" is found in many parts of Guiana as well as
" in Brazil, New Spain, and the large iflands.
" It has the fmell of carrion, which nothing can

* Though this bird refembles the turkey by its head, neck, and
the bulk of the body, it is by no means of that genus, but of that
of the Vulture; to which it is analogous by its inftincts, habits,
and, befides, by its hooked bill and its talons.

" remove;

" remove; the rump has been torn from it at
" the inftant of its being killed, and the entrails
" extracted, but all to no effect; for the flefh,
" which is hard, tough, and ftringy, ftill re-
" tained an infupportable odour. "Thefe birds,"
fays Kolben, " feed upon dead animals: I my-
" felf have often feen the fkeletons of cows,
" oxen, and wild beafts, which they had de-
" voured. I call thefe veftiges fkeletons, and
" not without reafon; fince the birds detach
" with fuch dexterity the flefh from the bones
" and the fkin, that what is left is a perfect
" fkeleton, covered ftill with the fkin, without
" the leaft derangement of the parts. One could
" hardly perceive that the carcafe is hollow till
" he is near it.—They perform it in this way:
" They firft make an opening in the belly of
" the animal and tear out the entrails, which
" they eat; they then enter the hollow and fe-
" parate the flefh. The Dutch of the Cape call
" thefe Eagles *Stront-vogels*, or *Stront-jagers;* that
" is, *dung birds.* It often happens that an ox,
" after being unyoked from the plough, and
" allowed to return alone to its ftall, lies down
" by the way to reft itfelf; and if thefe Eagles
" obferve its unguarded pofture, they infallibly
" fall upon it and devour it.—When they want
" to attack a cow or an ox, they collect to the
" number of a hundred or more, and pour at
" once upon the unfortunate victim. They
" have fo quick a fight, that they can difcern

K 4   " their

" their prey at an amazing height, and when
" it would escape the most acute eye ; and, when
" they perceive the favourable moment, they
" descend directly upon the animal, which they
" watch.   These Eagles   are   rather   larger
" than wild geese, their feathers are partly
" black, partly light grey, but the black pre-
" dominates; their beak is thick, hooked, and
" pointed ; their claws large and sharp."—"This
" bird," says Catesby, " weighs four pounds
" and a half ; the head and part of the neck
" is red, bald and fleshy as in the turkey, beset
" with straggling bristles ; the bill is two inches
" and an half long, partly covered with flesh,
" and its tip, which is white, is hooked like
" that of the falcon, but it has no whiskers at
" the sides of the upper mandible ; the nostrils
" are large and open, placed before at a remark-
" able distance from the eyes ; the plumage
" through the whole of the body has a mixture
" of deep purple and green ; its legs short and
" flesh-coloured, its toes long as in the domestic
" cocks, and its nails, which are black, are not
" so much hooked as those of falcons.   They
" feed on carrion, and fly continually on the
" search ; they continue long on the wing, and
" rise and descend so smoothly, that the motion
" of their pinions cannot be perceived.   A dead
" carcass attracts numbers of them ; and it is
" amusing to see their disputes with each other
                                    " while

" while eating *. An Eagle often prefides at
" the banquet, and does not fuffer them to
" approach till he has fatisfied his appetite.
" Thefe birds have a moft acute fcent, and fmell
" carrion at a vaft diftance, to which they re-
" fort from all quarters, wheeling about and
" making a gradual defcent till they reach the
" ground. It is generally fuppofed that they
" eat no living prey; but I know that fome of
" them have killed lambs, and that they common-
" ly feed on fnakes. They ufually rooft in num-
" bers together on old pines and cypreffes,
" where they continue feveral hours in the
" morning, their wings being difplayed †. They
" are very tame, and, while at their meals, will
" fuffer a very near approach."

I have thought proper to produce, at confiderable length, all the facts which tend to
throw light on the hiftory of this bird; for it
is in diftant countries, and efpecially in defert
regions, that we are to contemplate Nature in
her primæval form. Our quadrupeds, and even
our birds, perpetually driven from their haunts,
lofe in part their original inftincts, and acquire

---

* This fact is directly contrary to what Nieremberg, Marcgrave,
and Defmarchais afferts, with regard to the filence and concord
that prevail in their meals.

† This circumftance ftill farther fhews, that this bird belongs
to the Vultures; for when thefe fit they always keep their wings
fpread. Mr. Pennant fuppofes that they expofe their plumage to
the air, with the view of getting rid of the rank fœtor.

habits

habits which have a reference to the ftate of civil
fociety. We muft ftudy the difpofitions of the
Vultures in the folitary tracts in America, to difcover what would be the manners of our own,
if they were not molefted in their retreats,
checked in their multiplication, and difturbed in
their operations by our crowded population.—
Thefe are their primitive characters.—In every
part of the globe, they are voracious, flothful,
offenfive, and hateful; and, like the wolves, are
as noxious during their life, as ufelefs after
their death. [A]

[A] We may add from Mr. Pennant, that the Carrion Vultures
are not found in the northern regions of the Ancient Continent;
but in the New they are common through its whole extent, from
Nova Scotia to Terra del Fuego, and alfo in the Weft India iflands,
though they are faid to be fmaller there than on the main-land.
They fwarm in the torrid zone; and about Carthagena efpecially,
they haunt inhabited places, fit in numbers on the roofs of houfes,
or walk with fluggifh pace along the ftreets. They watch the
female alligator as fhe hides her eggs in the fand, and on her retreat, they hurry to the fpot, and eagerly lay bare the depofitory,
and devour the whole contents.

The CONDUR.

Le Condor, Buff.
Vultur Gryphus, Linn.
Avis ingens Chilensis *, *Cuntur*, Ray.

IF the power of flying conftitute the effential character of birds, the Condur Vulture muft be confidered the largeft of all.
The Oftrich, the Galeated Caffowary, and the
Hooded Dodo, whofe wings and feathers are
not calculated for flying, and who for this reafon
cannot quit the ground, ought not to be compared with it ; they are (if I may be allowed the
expreffion) imperfect birds, a fort of terreftrial
bipeds, which form an intermediate fhade between the birds and quadrupeds on the one hand ;
while the rouffette and rougette and the bats form
a fimilar fhade, on the other, between the quadrupeds and the birds. The Condur poffeffes, even in
a higher degree than the Eagle, all the qualities,
all the endowments which Nature has beftowed
on the moft perfect fpecies of this clafs of beings. Its wings extend eighteen feet ; the body,
the bill, and the talons are proportionally large
and ftrong; its courage is equal to its ftrength, &c.

* The Indians who live near the river Amazons, call it *Ouyrad-
Ouaffou*; that is, the Great *Ouara* or *Aura*.

—We

—We cannot give a better idea of its form, and
the proportions of the feveral parts of its body,
than by an extract from Father Feuillée, the
only naturalift and traveller who has given a
full defcription of this bird :—" The Condur is
" a bird of prey which haunts the valley of
" of Ylo in Peru.——I difcovered one that was
" perched upon a great rock: I approached it
" within mufket fhot, and fired, but, as my
" piece was only loaded with fwan-fhot, the
" lead could not pierce its feathers. I perceived
" however, from its motions, that it was wound-
" ed ; for it rofe heavily, and could with dif-
" ficulty reach another great rock five hundred
" paces diftant upon the fea-fhore. I therefore
" charged my piece with a bullet, and hit the
" bird under the throat. I then faw that I had
" fucceeded, and I ran to fecure the victim;
" but it ftruggled obftinately with death ; and
" refting upon its back, it repelled my attempts
" with its extended talons. I was at a lofs
" on what fide to lay hold of it; and I believe
" that if it had not been mortally wounded, I
" fhould have found great difficulty in fecuring
" it. At laft I dragged it down from the top
" of the rock, and, with the affiftance of a
" failor, I carried it to my tent, to figure it, and
" make a coloured drawing.

    " The wings of the Condur, which I meafured
" very exactly, extended eleven feet four inches
" from the one extremity to the other, and the
                                        " quill-

" quill-feathers, which were of a fine fhining
" black, were two feet two inches long; the thick-
" nefs of the bill was proportioned to that of the
" body, and its length was three inches and feven
" lines, the upper part pointed, hooked, and white
" at the end, and all the reft black ; a fmall fhort
" down of a tawny colour covered the whole head;
" the eyes were black, and encircled with a brown-
" ifh-red ring; the under-furface of its wings,
" and the lower part of its belly as far as the tail,
" were of a light brown; the upper furface of
" the fame colour, but darker; the thighs were
" covered as low as the knee with brown feathers;
" the *os femoris* was ten inches and a line in
" length, the tibia five inches and two lines; the
" foot was compofed of three anterior pounces and
" one pofterior, the laft being an inch and half
" long, with a fingle articulation, terminated by a
" black nail nine lines in length; the middle or
" great anterior pounce was five inches eight
" lines, with three articulations, and the nail
" which terminated it was an inch and nine lines,
" and was black like the others; the inner pounce
" was three inches two lines, with two articula-
" tions, and terminated by a nail of the fame fize
" with that of the great pounce; the outer pounce
" was three inches long, with four articulations
" and a nail of an inch; the tibia was covered
" with fmall black fcales, and fo were the pounces,
" only thefe were larger.

" Thefe animals commonly fettle upon the
" moun-

" mountains, where they procure their fubfift-
" ence; they refort to the fhore only in the
" rainy feafons; and feeling the approach of
" cold, they feek for fhelter and warmth in the
" plains. Thefe fummits, though fituated under
" the torrid zone, are yet expofed to a chill
" air; they are covered almoft the whole year
" with fnow, but particularly in winter, which
" had fet in on the 21ft of this month.

" The fcanty fubfiftence which thefe animals
" can pick up upon the margin of the fea, ex-
" cept when ftorms caft afhore large fifh,
" obliges them to make but a fhort ftay; they
" appear on the beach generally about evening,
" and there pafs the night, and return to their
" proper haunts in the morning."

Frezier, in his Voyage to the South Sea,
fpeaks of this bird in the following terms :—
" One day we killed a bird of prey, called
" *Condur*, whofe wings meafured nine feet; it
" had a brown comb, but not jagged like that
" of the cock; it had in the forepart a red
" gizzard, naked as in the turkey; it is com-
" monly bulky, and can with eafe carry
" off a lamb. Garcilaffo fays, that he found
" fome in Peru whofe wings extended fixteen
" feet."

In fact, it appears that thefe two Condurs,
mentioned by Feuillée and Frezier, were young
and uncommonly fmall for the fpecies; and
accordingly all the other travellers reprefent

I   them

them of a greater fize *. Fathers Abbeville and
Laët affirm, that the Condur is twice as large as
the eagle, and fo ftrong that it can pounce and
devour a whole fheep ; that it fpares not even
ftags, and eafily overthrows a man. There are
fome, fay Acofta and Garcilaffo †, whofe wings
extend fifteen or fixteen feet ; their beak is fo
firm, that they pierce a cow's hide, and two of
them are able to kill the animal, and eat the
carcafs. Sometimes they even dare to attack
men ; but fortunately they are rare, for if they
were numerous, they would extirpate the cattle.
Defmarchais relates that thefe birds meafure
eighteen feet acrofs the wings ; that their talons
are thick, ftrong, and very hooked ; that the
American Indians affirm, that they feize and
tranfport a hind or a young cow as they would
do a rabbit ; that they are of the fize of a fheep,

* On the coaft of Chili (fays Strong), not far from the ifland of
Mocha, we met with this bird (the Condur) fitting on a lofty cliff
nigh the fhore. We fhot it, and the failors were aftonifhed at its
prodigious bulk, for its wings meafured from tip to tip thirteen
feet. The Spaniards fettled in that country told us, that they
dreaded left thefe birds fhould carry off and devour their children.
RAY, Synop. Av.

† Thofe who have meafured the Condurs have found that their
wings extend fixteen feet. Their bill is fo ftrong and fo hard that
they eafily pierce an ox's hide. Two of thefe birds attack a cow
or a bull, and fucceed. They can pounce children of ten or twelve
years old, and prey upon them. Their plumage is like that of
magpies ; they have a comb on the forehead, which is different
from that of cocks, not being jagged ; their flight is terrible, and
when they alight on the ground one is ftunned with the noife of
the ruftle. Hift. Incas.

and

and that their flesh is coriaceous, and smells like carrion; that their sight is sharp, their aspect stern and cruel; that they seldom frequent the forests, where they have scarcely room to wield their enormous wings; but that they haunt the sea-shore, the sides of rivers, and the savannahs, or natural meadows *.

Ray, and almost all the naturalists after him, have considered the Condur † as a kind of Vulture, because its head and neck are destitute of plumage. But there is still reason to doubt this conclusion; for it appears that its dispositions have a greater resemblance to those of the Eagles. It is, say the travellers, bold and intrepid; it ventures alone to attack a man, and can, with little difficulty, kill a child ten or twelve years old ‡. It stops a whole flock of
sheep,

* " Our sailors," says Spilberg, " caught in the island of Loubet, on the coast of Peru, two birds of an uncommon size, having bills, wings and talons like the Eagles, a neck like the sheep, and a head like the cock. Indeed their figure was as extraordinary as their bulk."

" There were," says de Solis, " in the *menagerie* of the Emperor of Mexico, birds of such an astonishing size and ferocity, that they seemed to be monsters. . . . Their voracity was prodigious; and an author mentions, that one of them devoured a sheep at each meal."

† To this genus the large bird of Chili called *Condur* seems to belong; I have been able, from this imperfect description, to come to this conclusion, since I cannot doubt that it is a Vulture, being named *Aura*. On account of its naked head, it was at first supposed by the sailors to be a turkey-cock.—From a similar inadvertency our first American colonists imagined the Carrion Vulture to be a turkey.

‡ " It has often happened that one of these birds has killed and eaten children of ten or twelve years old." SLOANE, *Phil. Transf.*
" The

sheep, and, at its leisure, selects its prize. It
carries off roebucks, kills hinds and cows ; and
also catches large fish. It therefore lives, like
the eagle, upon the ravages which it commits ;
it feeds upon fresh prey, and not upon dead
carcasses.—These are rather the habits of the
eagle than of the vulture.

It appears to me that this bird, which is still
but little known because it is very rare, is not
confined to South America ; I am confident
that it is found both in Africa and Asia, and
perhaps even in Europe. Garcilasso properly
remarks, that the Condur of Peru and of Chili,
is the same bird with the *Ruch* or *Roc*, of the
eastern nations, so famous in the Arabian
Tales, and which is mentioned by Marco Paolo ;
and, with equal propriety, he quotes Marco
Paolo along with the Arabian Tales ; for, in the
account of the Venetian, there is almost as
much exaggeration. " In the island of Mada-
" gascar," says he, " there is a wonderful kind
" of bird, which they call *Roc*, which bears a

" The famous bird called the *Cuntur* in Peru, which I have seen
in several parts among the mountains of Quito, is also found, if I
am rightly informed, in the low-lands near the river Maragnon.
I have seen it hovering over a flock of sheep ; it is probable that
the presence of the shepherd prevented its attack. It is an opinion
universally entertained, that this bird can bear off a roebuck, and
sometimes preys on a child. It is said that the Indians decoy it,
by presenting the figure of a child formed of a very viscous clay,
upon which he darts with rapid flight, and impresses his claws so
deeply, that he cannot disentangle himself." *Voyage de la Riviere
des Amazons, par M. de la Condamine.*

" refemblance to the eagle, but is incompa-
" rably bigger . . . . . the wing-feathers being
" fix fathoms long, and the body large in pro-
" portion. Its ftrength is fo aftonifhing, that,
" fingly and unaffifted, it feizes an elephant,
" hurries the ponderous animal into the air,
" drops it, and kills it by the fall, and after-
" wards feeds upon the carcafe." It is unneceffary to make any critical reflections; it is
fufficient to oppofe facts of greater veracity,
fuch as we have already related, and what we
fhall ftill produce. It appears to me that the
bird mentioned almoft as large as an Oftrich in
the Hiftory of the Voyage to the Southern
continent *, which the Prefident de Broffes has
digefted with as much judgment as care, muft
be the fame with the Condur of the Ameri-
cans, and the Roc of the Orientals. Moreover,
I am of opinion, that the bird of prey found in
the vicinity of Tarnafar †, a city in the Eaft
Indies,

* " From the boughs of a tree which produces the fruits called
Monkey's bread, were fufpended nefts that refembled large oval ham-
pers, open below, and loofely interwoven with branches. I had not
the fatisfaction to fee the bird to which thefe belonged; but the
people in the neighbourhood affured me, that its figure was much
like that of the kind of eagle which they call *Ntann*. 'To judge of
the bulk of thefe birds from that of their nefts, it cannot be *much
inferior to that of the Oftrich."* *Hift. des Navigations aux Terres
Auftrales.*

† " In the vicinity of Tarnafar, a city of India, are many kinds
of birds, fubfifting chiefly on prey, and much larger than the eagles;
for the hilts of fwords are formed of the upper part of the bill; that
part of the bill is fulvous, varied with cœrulean; but the colour of
the

Indies, which is much larger than the eagle,
and whose bill serves for the hilt of a sword, is
likewise the Condur; as well as the Vulture of
Senegal*, which attacks and carries off children;
and that the savage bird of Lapland †, as large
and thick as a sheep, mentioned by Regnard
and Martiniere, and of whose nest Olaus Magnus gives an engraving, is probably the same.
But not to draw our comparisons from such
distant countries, to what other species must we
refer the *Laemmer Geyer (Lamb-Vulture)* of the
Germans ‡ ? This Vulture, the plunderer of
lambs and sheep, which has been often seen at
different

the bird is black, with a few straggling feathers of purple." *Lud.
Patricius apud Gesnerum.*

* There are in Senegal Vultures as large as eagles, which devour young children when they find them alone. *Voyage de la
Maire.*

† There is found also in Russian Lapland a wild bird of a pearl
grey, as thick and large as a sheep, having a head like a cat, and
eyes glaring and red; the bill of an eagle, and the feet and talons
of the same. *Voyage des Pays Septentrionaux, par de la Martinere.*

There are scarcely fewer birds than quadrupeds in Lapland; the
eagles are to be met with in abundance; some are so prodigiously
large that, as I have already said, they seize the young fawns of the
rein-deer, and carry them to their nests, which they construct on the
summit of the highest trees; and for this reason some person is always set to watch these. *Regnard, Voyage de Lappon.*

‡ It may be proper to observe, that the *Laëmmergeyer* has been
since discovered to be quite a different bird from the Condur. It is
the *Vulture-Eagle* of Albin, the *Vultur Barbatus* of Linnæus, and
the *Falco Barbatus* of Gmelin: It is the same with the *Avoltoio
Barbato* of the Italians, the *Alpine Vulture* of Brisson, and the
Percnopterus Gypaëtos of Ray and Willoughby. " It is whitish
" flame-coloured; the back dusky, with a black stripe above and

different times in Germany and Switzerland, and which is much larger than the agle, muſt be the Condur. Geſner relates, from the teſtimony of an author of credit (George Fabricius) the following facts. Some peaſants between Mieſen and Briſa, cities in Germany, loſing every day ſome of their cattle, which they in vain ſought for in the foreſts, obſerved a very large neſt reſting on three oaks, conſtructed with ſticks and branches of trees, and as wide as would cover a cart. They found in this neſt three young birds already ſo large, that their wings extended ſeven ells; their legs were as thick as thoſe of a lion, the nails of the ſize of a man's fingers; and in the neſt, were ſeveral ſkins of calves and ſheep. Valmont de Bomare and Salerne have thought, as well as myſelf, that the *Laemmer Geyer* of the Alps muſt be the Condur of Peru. Its ſpread wings, ſays Bomare, extend fourteen feet; it commits dread-

"below the eyes." It inhabits in ſmall flocks the Alpine tracts of Switzerland, and of the country of the Griſons. It neſtles in the holes of inacceſſible rocks. It reſembles the Vulture in its exterior appearance, in its gregarious habits, and in its fondneſs for carrion. It is like the eagle in its head and neck, and in its courage.

To the ſame ſpecies belong two varieties which are found in the mountains of Perſia. The firſt is the *Golden Vulture* of Briſſon and Latham, and the *Cheſnut Vulture* of Willoughby. "It is rufous; "the back black; the head, and the under part of the neck, of a "tawny white; the wings and tail-feathers duſky."—The ſecond is the *Falco Magnus* of Gmelin. "Its cere is cœrulean; its feet, "and the under part of its body, are cheſnut mixed with white; "its tail cinereous."

ful

ful havoc among the goats, the sheep, the cha-
mois, the hares and the marmots. Salerne also
relates a decisive fact on this subject, which deserves to be quoted at length. " In 1719, M. Dé-
" radin, father-in-law to M. de Lac, shot at his
" castle of Mylourdin, in the parish of Saint-
" Martïn d'Abat, a bird which weighed eighteen
" pounds, and whose wings measured eighteen
" feet. It flew for some days about a pond,
" and was pierced by two balls under the wing.
" The upper part of its body was mottled with
" black, grey, and white ; the top of its belly
" red as scarlet ; and its feathers were crisped.
" They ate of it both at the castle of Mylour-
" din and at Châteauneuf-sur-Loire ; the flesh
" was found tough and fibrous, and smelt
" of the marsh. I saw and examined one of
" the small feathers of the wings, which was
" larger than the quill-feather of the swan.
" This singular bird seemed to be the Condur."
In short, the enormous size must be considered
as a decisive character ; and though the *Laemmer Geyer* of the Alps differs from the Condur
of Peru by the tints of its plumage, we cannot
but refer them to the same species, at least till
we obtain a more accurate description of both.

It appears from the testimonies of travellers,
that the Condur of Peru has a plumage marked
with black and white, like that of the magpie ;
and therefore the large bird killed in France at
the castle of Mylourdin resembles it both in size

L 3   and

and colour. We may therefore conclude, with great probability, that thefe exalted fort of birds, though very rare, are fcattered over both continents; and feeding upon every kind of prey, and dreading nothing but the human race, avoid the habitations of men, and confine their haunts to extenfive deferts, or lofty mountains.

THE KITE

The KITE and the BUZZARDS.

THESE ignoble, filthy, and flothful birds ought
to follow the vultures, which they refemble by their difpofitions and habits. The vultures, though deftitute of every generous quality, claim, by their fize and ftrength, a principal rank among the feathered race. The Kites
and Buzzards, inferior to thefe in force and
magnitude, far exceed them in numbers. They
are more troublefome ; they oftener vifit inhabited fpots, and fettle nearer the refidence of
men ; they build their nefts in places more ac-
ceffible ; they feldom remain in deferts, but prefer the fertile hills and dales to the barren mountains. In fuch fituations, Nature, affifted by the
forming induftry of man, teems with vegetable
and animal productions, and prefents an eafy
and abundant harveft to the voracious tribe.
The Kites and Buzzards are neither bold nor
timid ; they have a kind of ftupid ferocity, which
gives them an air of cool intrepidity, and feems
to remove the fenfe of danger. It is eafier to
approach and to kill them than the eagles or
vultures ; when detained in captivity, they are
lefs capable of inftruction ; and they have always

been

been profcribed and erafed from the catalogue of
noble birds, and banifhed from the fchool of
falconry. In all ages, it has been common to
compare a grofs fhamelefs man to a Kite, and
a difgufting ftupid woman to a Buzzard.

Though thefe birds refemble each other in
their inftinct, their fize *, and the form of their
bill and other characters, the Kite is however
eafily diftinguifhed, not only from the Buzzards,
but from all other birds of prey, by a fingle
prominent feature : its tail is forked ; the middle
feathers being fhorter than the reft, leave a blank
which can be perceived at a diftance, and which
has improperly given occafion to the name of
Forked-tail-Eagle. Its wings are alfo proportionally longer than thofe of the Buzzard, and it
flies with far greater eafe. It fpends its life in
the region of the clouds ; it feldom courts repofe,
and every day it traverfes an immenfe range.
But it performs thefe continual circling journies,
not with the view to procure its prey ; it only
indulges its natural, its favourite exercife. One
cannot but admire the eafe and the elegance of
its motion ; its long narrow wings feem perfectly
fixed ; the tail alone appears to direct all its evo-
lutions, and it quivers inceffantly ; it rifes without making an exertion, and defcends as if it

* The Royal Kite is in fize and figure like the Buzzard.——
In the former, the legs are faffron colour, and fhorter ; and in the
Buzzard, the part below the knee is covered with pendent ferru-
ginous feathers. SCHWENCKFELD.

were

were gliding along an inclined plane; it accelerates its course, it retards it; it stops, hovers suspended in the same place for whole hours, nor is observed even in the least to quiver its expanded wings.

In our climate, there is only one species of Kite, which the French call the *Royal Kite* *, because it was formerly an amusement for princes to hunt this cowardly bird with the falcon or the sparrow-hawk. It is indeed entertaining to see it, though possessed of all that ought to inspire courage, and deficient neither in weapons, strength, nor agility, decline the combat, and fly before a sparrow-hawk smaller than itself; it constantly circles, and rises, as it were, to conceal itself in the clouds, and when overtaken, it suffers itself to be beaten without resistance, and brought to the ground, not wounded, but vanquished, and rather overcome with fear, than subdued by the force of its antagonist.

* *Le Milan Royal*, Buff. *Falco Fulvus*, Linn. *The Kite* or *Glead*, Will.

In German it is named *Woike*, *Weisser Milan*, *(White Kite,)* and *Hungeyer (Hen-Vulture)*: in Dutch, *Wowe*: in Polish, *Kania*: in Swedish, *Glada*: in Spanish, *Milano*: in Italian, *Milvio*, *Nibbio*, *Poyana*. The antient Greeks called it Ἰκτιν, a word which is also employed to denote a sort of hare. (Buffon supposes, probably from an oversight, the term to be Ἰκτις, which signifies a polecat, and imagines that it was applied to the Kites, because poultry is the common prey of both.) The Romans named it *Milvius*, i. e. *Mollis Avis*, indolent bird. The Swedish, *Glada*, and the Old English name *Gleade*, refer to its gliding motion.

Though

Though the Kite fcarcely weighs two pounds
and an half, and meafures only fixteen or
feventeen inches from the point of the bill to
the toes, its wings extend near five feet; the
cere, the iris, and the feet are yellow; the bill
is of a horn colour, blackifh towards the point,
and the nails are black; its fight is as keen
as its flight is rapid; fometimes it foars fo high
in the air, as to be beyond the reach of our view,
and yet at this immenfe diftance, it diftinctly
perceives its food, and defcends upon whatever
it can devour or ravage without refiftance; its
attacks are confined to the fmalleft animals and
the feebleft birds; it is particularly fond of young
chickens, but the fury of the mother is alone
fufficient to repel the robber. " Kites," one
of my friends writes me *, " are exceedingly
" cowardly animals. I have feen two of them
" chafe a bird of prey together to fhare the
" fpoils, and yet not fucceed. The ravens in-
" fult them, and drive them away. They are
" as voracious as they are daftardly; I have feen
" them pick up, on the furface of the water,
" fmall dead and half rotten fifh; I have ob-
" ferved fome carry a large viper in their claws;
" others feed upon the carcaffes of horfes and
" oxen. I have feen fome alight upon tripes,
" which women were wafhing by the fide of
" a rivulet, and fnatch it almoft in their prefence.

* Mr. Hebert, to whom I am indebted for feveral important
facts with regard to the hiftory of Birds.

I   " I once

" I once offered a young Kite, which the chil-
" dren were breeding in the house where I
" lived, a pretty large young pigeon, and it
" swallowed it entire with the feathers."

This sort of Kite is common in France, especially in the provinces of Franche-compté, Dau-
phiné, Bugey, Auvergne, and in all the others
which are in the vicinity of mountains. It is
not a bird of passage, for it constructs its nest
in these countries, and breeds in the hollow of
rocks. It appears even that they nestle in
England, and remain there during the whole
year *. The female lays two or three eggs,
which are whitish, with pale yellow spots, and
like those of all the carnivorous birds, are rounder
than hen's eggs. Some authors have said that
they build their nests in the forests, upon old oaks
or firs. Without venturing absolutely to deny
the fact, we can affirm that they are commonly
found in the holes of rocks.

This species seems to be scattered through the
whole extent of the ancient continent, from
Sweden to Senegal †; but I am uncertain if it
be

---

* Some have supposed these to be birds of passage; but
in England they certainly continue the whole year. *British
Zoology.*

† It appears that the Kite is found in the north; since Linnæus
includes it in his catalogue of the Swedish birds, describing it as
*a falcon with a yellow cere, forked tail, ferruginous body, and whitish
coloured head.* Travellers also tell us, that it occurs in the hottest
parts of Africa. In Guinea, says Bosman, the Kites not only
plunder hens, from which circumstance they have their name, but
whatever

be alfo found in the new; for no mention is made of it in the accounts that are given of America. There is indeed a bird, which is faid to be a native of Peru, and appears in Carolina only in fummer, which in fome refpects refembles the Kite, and has like it a forked tail. Catefby gives a defcription and figure of it under the name of *Swallow-tailed-Falcon*, and Briffon terms it the *Carolina Kite* [A]. I am inclined to believe that it is a fpecies related to that of our Kite, and occupies its place in the new continent.

But there is another fpecies ftill nearer related, and which appears in our latitudes as a bird of paffage; it is called the *Black Kite*. Ariftotle diftinguifhes this from the preceding, which he

whatever they can difcover and feize, whether game or fifh; and their audacity is fo great, that they fnatch the fifh from the negro girls, which they carry to market, or call in the ftreets. *Voyage to Guinea.*

Near the defert bordering Senegal, fays another traveller, there is a fort of Kite.—Every thing fuits its greedy appetite; it is not intimidated by fire-arms. Flefh, whether raw or dreffed, tempts it fo ftrongly, that it feizes the morfels as the failors convey them to their mouth. *Hift. Gen. des Voyages par M. Abbé Prevoft.*

[A] The *Swallow-tailed-Falcon* is the *Falco Furcatus* of Linnæus, the *Great Peruvian Swallow* of Feuillé, and the *forked-tailed Peruvian Falcon* of Klein. The fpecific character:—" The cere is " dull coloured, the feet yellowifh, the body dufky above and " whitifh below, its tail very long and forked." It inhabits Carolina and Peru, lives upon infects and ferpents, and is migratory. It is rather fmaller than the Kite, but of the fame length. The *irides* are red, the head and neck fnowy, the back dufky or black, fhining with purple or green.

names

names fimply *Kite*, and gives it the epithet of
Ætolian, becaufe, in his time, it was probably
moft common in Ætolia *. Belon alfo mentions
thefe two Kites; but he is miftaken when he
fays that the firft, which is the Royal Kite, is
blacker than the fecond, which he notwithftanding calls *Black Kite*. Perhaps it is an error of
the prefs, for it is certain that the Royal Eagle
is not fo black as the other. No naturalift, ancient or modern, has attended to the moft obvious
diftinction between thefe two birds, which is
founded in the difference of the figure of their
tails. But in fize, their fhape and their inftinctive habits, they bear a clofe refemblance,
and muft therefore be confidered as kindred
fpecies †.

Aldrovàndus fays, that the Hollanders call this
Kite *Kukenduff*, and that though fmaller than
the Royal Eagle, it is ftronger and more agile.
Schwenckfeld affirms on the contrary, that it
is weaker and more flothful, and that it preys
only upon field-mice, grafshoppers, and upon
fmall birds, as they rife from their nefts.

* Kites have for the moft part two eggs, fometimes three ;
and they hatch as many young. But what is called the Ætolian
lays fometimes four. *Arift. Hift. An.*

† The *Falco Ater* of Gmelin, the *Schwartzer Milan*, and the
Brauner Mald Geyer of the Germans, and the *Black Kite* of Sib-
bald and Latham. "Its cere and feet are yellow, its body dufky-
"black above, whitifh on the head and the under part of the
"body, its tail forked." It is fmaller than the common kite, its
tail flightly forked, its legs flender ; its egg is ferruginous, with
dufky and black fpots.

He

He adds, that this fpecies is very common in Germany; this may be true, but we are certain that in France and England it is much lefs frequent than the Royal Kite. The one is a native, and refides with us the whole year; the other is a bird of paffage, which quits our climate in autumn, and migrates to warmer regions. Belon was an eye-witnefs of their paffage from Europe to Egypt;—before the approach of winter, they traverfe the Black-fea, marfhalled in numerous lines, and return in the fame order about the beginning of April : they remain the whole winter in Egypt, and are fo tame, that they enter the cities, and alight upon the windows of the houfes; their fight is fo quick, and their flight fo fteady, that they catch in the air the bits of meat that are thrown to them [A].

[A] Mr. Pennant reprefents the Kite as larger than is ftated by the Count ; it weighs forty-four ounces, and is twenty-feven inches long. It inhabits England in all feafons.

THE COMMON BUZZARD

The BUZZARD*.

La Buse, Buff.
Falco Buteo, Linn.
Goiran, Hift. de Lyons.
Maaffe Geyer, Gunth.
Pojana Secunda, Zinn.

THE Buzzard is fo common and fo well
known, that it requires no particular defcription. Its length is twenty or twenty-
one inches; its alar extent four feet and an
half; its tail is only eight inches, and the
wings, when clofed, reach a little beyond its
point; the iris is of a pale yellow, and almoft whitifh; the cere and feet are yellow, and
the nails black.

This bird refides the whole year in our forefts;
it appears ftupid, whether in the domeftic ftate,
or in that of liberty; it is fedentary, and even
indolent; it often continues for feveral hours
together perched upon the fame tree; its neft is
conftructed with fmall branches lined in the infide with wool, and other foft, light materials;
it lays two or three eggs, which are whitifh,

---

* In Italian, it is called *Buzza* or *Bucciario*. The Latin name
is *Buteo*; the Greek Τριοχης, probably from the notion that it had
three tefticles τρεις and ορχις.

fpotted

ſpotted with yellow. It feeds and tends its young
longer than the other birds of prey, moſt of
which expel their brood before they are able to
provide with eaſe for themſelves. Ray even affirms that if the mother happen to be killed in
the time of her tender charge, the male Buzzard
patiently diſcharges the truſt.

This bird of rapine does not ſeize its prey on
the wing; it ſits on a tree, a buſh, or a hillock,
and darts upon the humble game that comes
within its reach. It catches young hares and
young rabbits, as well as partridges and quails; it
commits havoc upon the neſts of moſt birds;
and when more generous ſubſiſtence is ſcanty,
it ſubſiſts upon frogs, lizards, ſerpents, and
graſs-hoppers.

This ſpecies is ſubject to great variety, ſo that
if we compare five or ſix common Buzzards
together, we ſhall ſcarcely find two that are
alike. Some are entirely white; in others, the
head only is white; others again are mottled
with brown and white. Theſe differences are
òwing chiefly to the age and ſex, for they are all
found in our own climate [A].

[A] The ſpecific character of the Buzzard, *Falco Buteo*, LINN.
—" Its cere and feet are yellowiſh, the body duſky, the abdomen
" pale, with dirty ſpots, the tail ſtreaked with duſky colours."

The HONEY BUZZARD.

La Bondrée, Buff.
Falco Apivorus, Linn.
Pojana, Zinn.

AS the Honey Buzzard differs but little from
the Common Buzzard, they have been diftinguifhed by thofe only who have carefully compared them. They have indeed more analogous
than difcriminating characters, but the difference
of external appearance and of natural habits, is
fufficient to conftitute two fpecies; which, though
allied, are yet feparate and independent. The
Honey Buzzard is as large as the Buzzard, and
weighs near two pounds; its length from the
point of the bill to the end of the tail is twenty-
two inches; its wings extend four feet two
inches, and when clofed reach to three-fourths
of the tail; its bill is rather longer than that of
the Buzzard; the cere is yellow, thick, and unequal *; its noftrils are long and curved; when
the bill opens, the mouth appears very large and
of a yellow colour; the iris is of a bright yellow; the legs and feet are of the fame colour,

* Some naturalifts have faid that the bill is black; but we may
prefume that this difference is owing to age, fince it is firft white;
perhaps it becomes fucceffively yellow, brown, and black.

VOL. I.　　　　　M　　　　　and

and the nails, which are not much hooked, are ſtrong and blackiſh; the head is large and flat, and of a grey cinereous. A full deſcription of this bird occurs in the work of Briſſon and in that of Albin; the laſt author, after deſcribing the external parts of the Honey Buzzard, ſays, that its inteſtines are ſhorter than in the Common Buzzard; and he adds, that there are found in the ſtomach of the Honey Buzzard ſeveral green caterpillars, as alſo ſome common caterpillars and other inſects.

Theſe birds, as well as the Common Buzzards, build their neſts with ſmall ſticks, and line them with wool; their eggs are of an aſh-colour, and marked with ſmall brown ſpots. Sometimes they take poſſeſſion of the neſts of other birds; for they have been found in an old neſt of the Kite. They feed their young with cryſalids, and particularly with thoſe of waſps. The heads and different portions of waſps have been diſcovered in a neſt in which were two young Honey Buzzards. At that tender age, they are covered with a white down, ſpotted with black; the feet are of a pale yellow, and the cere white. In the ſtomach of theſe birds, which is very large, there are alſo found entire frogs and lizards. The female is larger than the male, and both trip and run, without the aſſiſtance of their wings, as faſt as our dunghill cocks.

Though Belon ſays that there is not a young ſhepherd in Limagne in Auvergne, who is not

acquainted

acquainted with the Honey Buzzard, and could not catch it with a fnare baited with frogs, or bird-lime, or even with a noofe, it is certain that at prefent they are more rare in France than the Common Buzzard. Among twenty Buzzards brought to me at different times in Burgundy, there was not a fingle Honey Buzzard ; nor do I know from what province the fpecimen came which we have in the king's cabinet. Salerne fays, that in the country of Orleans, the Common Buzzard is named Honey Buzzard ; but thefe may be different birds.

The Honey Buzzard generally fits upon low trees to fpy its prey. It catches field mice, frogs, lizards, caterpillars, and other infects. It fcarcely flies but from tree to tree, or from bufh to bufh, always low; nor does it foar like the Kite, which it refembles by its inftincts, but from which it can be readily diftinguifhed by its motions and the fhape of its tail. It is common to place fnares for the Honey Buzzard, becaufe in winter it is very fat and delicate to eat.

The BIRD SAINT MARTIN.

L'Oiseau Saint Martin, Buff. and Bel.
Falco Cyaneus, Linn.
Falco Torquatus (Mas.) Briff.
Pygargus Accipiter (Mas.) Ray and Will.
Falco Albanella, Gerin.
Lanarius Cinereus, Frif.
Falco Plumbeus Cauda Teffelata, Klein.
Hen Harrier, Penn. Edw. Will. Alb. and Lath.

T H E modern naturalifts have given this bird
the name of *Lanner Falcon* or *Cinereous
Lanner;* but it appears to be of a different genus
from either the Falcon or the Lanner. It is
rather larger than the common crow, and its
body is proportionally more delicate and flexible. Its legs are long and flender: whereas
thofe of the Falcon are robuft and fhort; and
the Lanner is defcribed by Belon to be ftill lower
on its legs than any Falcon; but in this charaƈter
it refembles the White John and the Ring-tail.
The only analogy therefore which fubfifts between it and the Lanner, is founded in the habit
of tearing with its bill all the fmall animals
which it catches, and in not fwallowing them
entire like the other large birds of prey. It
ought, fays Edwards, to be claffed with the Long-
tailed Falcons: in my opinion, it fhould be
ranged

ranged with the Buzzards, or rather placed next the Ring-tail, to which it is fimilar in its inftincts, and in many of its properties *.

This bird is pretty common in France, as well as Germany and England; the fpecimen which we have figured was killed in Burgundy. Frifch has given two plates of this fame bird, Nº 79 and 80, which differ fo little from each other, that we cannot confider them, as he does, a different fpecies; for the varieties which he remarks are fo flight, that they muft be afcribed folely to age or fex. Edwards, who alfo prefents an engraving of this bird, fays, that the fpecimen from which it was taken was killed near London; and he adds, that it was obferved to flutter about the foot of fome old trees, and fometimes to ftrike the trunks with its bill and claws, and that the reafon of the motion could not be perceived till after its death, when the body being opened, there were found in its ftomach twenty fmall lizzards, torn or cut into two or three portions.

* Belon does not hefitate to fay, that it is of the fame fpecies with the White John, and at the fame time he admits, that it is much like the Kite. "There is ftill another fpecies of the White "John or St. Martin, called *White-tail*, of the fame kind with "the abovefaid; but in colour it is much more like the Royal "Kite, though more flender—It refembles the Royal Kite fo much, "that we could not perceive the difference, were it not fmaller "and whiter under the belly, the feathers on the rump being "white both above and below, and hence it is named *White-* "*tail*."

When

When we compare this bird with what Belon fays of this fecond Saint Martin, we cannot doubt of their identity; and befides the refemblance in point of fize, fhape, and colour, their natural habits of flying low, and fearching eagerly and inceffantly for fmall reptiles, belong not fo much to the Falcons and other noble birds, as to the Buzzard, the Harpy, and others which partake of the groveling manners of the Kites. This bird, which is well defcribed and figured by Edwards, is different from what the authors of the Britifh Zoology name the *Hen Harrier.* Thefe are diftinct birds, of which the firft, what we call after Belon the *Saint Martin*, has, as I have faid, been mentioned by Frifch and Briffon under the name of *Lanner-Falcon* or *Cinereous Lanner;* the fecond, which is the *Subbuteo* of Gefner, and which we term *Soubufe*, has been named *White-tail-Eagle* by Albin, and *Collared-Falcon* by Briffon. Befides, the falconers call this bird Saint Martin, the *Hawk Harpy*. *Harpy* is among them a generic name, which they apply not only to the bird Saint Martin, but to the Ring-tail and the Red Buzzard, of which we fhall afterwards fpeak.

THE RINGTAIL FALCON.

The SOUBUSE.

La Soubuse, Buff.
Falco Pygangus, Linn.
Falco Torquatus (fem.) Briss.
Ring-tail, Penn. Alb. Will. Lath. &c.

THIS bird refembles the Saint Martin in its inftincts and habits; both fly low to catch field-mice and reptiles; both enter the outer-court, and haunt the places where poultry is kept, to feize young pigeons and chickens; both are ignoble birds, which attack only the weak and feeble, and therefore deferve neither the name of Falcons nor that of Lanners.

The male, as in other rapacious birds, is much fmaller than the female, and is befides diftin-guifhed by the want of a collar, that is of fmall feathers briftled round the neck. This differ-ence feemed to conftitute a fpecific character; but very fkilful falconers affured us, that it was only fexual; and upon examining more clofely, we found the fame proportions between the tail and the wings, the fame diftribution of colours, the fame form of the neck, head, and bill, &c. fo that we could not oppofe their opinion. But what occafioned more difficulty was, that almoft all the naturalifts have given the Ring-tail a different male, which is what we have named

M 4   Saint

Saint Martin ; and it was only after numberlefs
comparifons that we determined to fet afide
their authority. We fhall remark that the Soubufe is found both in France and in England ;
that it has long and flender legs like the Saint
Martin ; that it builds its neft in thick bufhes,
and lays three or four reddifh eggs ; and that
thefe two birds, with the one which we fhall
mention in the next article by the name of
Harpy, form a fmall genus more nearly allied
to the Kites and Buzzards than to the Falcons. [A]

[A] Both this and the preceding article are involved in obfcurity, arifing from the oppofite opinion of naturalifts ; fome main-
taining that the former is only the male of the latter, while others
confider them as of different fpecies.   The Saint Martin, or Hen-
harrier *(Falco Cyaneus,* LINN.) is thus charaƈerifed :—" Its cere
" is white ; its feet tawny ; its body of a hoary cœrulean, and a
" white arch above the eyes encircling the gullet." The Soubufe
or Ring-tail *(Falco Pygargus,* LINN.) :—" Its cere and feet are
" yellow ; its body cinereous ; its lower-belly palifh, with rufous
" oblong fpots ; its orbits white."   To throw greater light upon
the fubjeƈt we fhall borrow the account given in the Britifh Zoology.

" The male, or the Hen-harrier, weighs about twelve ounces ;
the length is feventeen inches ; the breadth three feet three inches ;
the bill is black ; cere, irides, and the edges of the eye-lids, yellow ; the head, neck, and coverts of the wings, are of a bluifh grey ;
the back of the head white, fpotted with a pale brown ; the breaft,
belly, and thighs, are white ; the former marked with a few fmall
dufky ftreaks : the fcapular feathers are of a deep grey, inclining
to dufky ; the two middle feathers of the tail are entirely grey ; the
others only on their exterior webs ; the interior being white, marked
with fome dufky bars ; the legs are yellow, long, and flender.

" The female weighs fixteen ounces ; is twenty inches long ; and
three feet nine inches broad : on the hind part of the head, round
the ears to the chin, is a wreath of fhort ftiff feathers of a dufky hue,

tips

tipt with a reddish white; on the top of the head and the cheeks the feathers are dusky, bordered with rust colour; under each eye is a white spot; the back is dusky; the rump white, with oblong yellowish spots on each shaft; the tail is long, and marked with alternate bars of dusky and tawny, of which the dusky bars are the broadest; the breast and belly are of a yellowish brown, with a cast of red, and marked with oblong dusky spots."

The HARPY*.

La Harpaye, Buff.
Circus Rufus, Gmel.
Fiſch-Geyer, Brand-Geyer, Friſch.
Harpy Falcon, Lath.

HARPY is an ancient generic name which has been beſtowed upon the Moor or Marſh Buzzards, and ſome other kindred tribes; ſuch as the Ringtail and the Hen-harrier, which has been termed the *Hawk Harpy.* We have conſidered the name as ſpecific, and have applied it to the bird which falconers at preſent call *Red-harpy,* and which our nomenclators term *Red-buzzard ;* and Friſch, improperly, *Middle Lanner Vulture,* as he has alſo improperly termed the Marſh Buzzard, *Great Lanner Vulture.* We have preferred the ſimple name of *Harpy,* becauſe it is certain that this bird is neither a Vulture nor a Buzzard. Its habits are the ſame with thoſe of the two birds which we have treated in the two preceding articles. It catches fiſh like the White John, and draws them alive out of the water. It ſeems, ſays

---

* It is diſtinguiſhed by " its yellow feet, its rufous body varie-
" gated with longitudinal ſpots, its back being duſky, and the fea-
" thers of the tail cinereous." The irides are ſaffron-coloured. Its
length is twenty inches.

Friſch,

Frifch, to have a more acute fight than any of the other birds of rapine, its eye-brows being more projected. It is found both in France and in Germany, and loves to haunt the fides of rivers and pools. In its inftincts it refembles the preceding, and therefore we fhall not be more particular.

The MOOR BUZZARD*.

Le Buſard, Buff.
Falco Æruginoſus, Linn. Gmel. Ray, Will. Klein, & Friſch.
Falco Bœticus, Gerin.
Faux-Perdrieux, Belon.
Il Bozzargo, Cet.
Il Nibbio, Zinn.

THIS bird was formerly called the *Baſtard Partridge*, and ſome falconers term it the *White-headed-Harpy*. It is more voracious and leſs ſluggiſh than the Common Buzzard; and this is perhaps the only reaſon that it appears not ſo ſtupid and more vicious. It commits dreadful havoc among the rabbits, and is equally deſtructive to the fiſh as to the game. Inſtead of haunting, like the Common Buzzard, the mountain-foreſts, it lodges only in the buſhes, the hedges, and ruſhes near pools, marſhes, and rivers that abound with fiſh. It builds its neſt at

---

* The Greek name is Κιρκος, whence the Latin *Circus*. In French it is commonly termed the *Marſh Buſard*; but as in that country there exiſts only one ſpecies of it, Buffon preſerves the ſimple name of *Buſard*. " Its cere is greeniſh; its body grey; its " crown, throat, *axillæ*, and feet, are yellowiſh." It varies extremely in regard to colour: in ſome the body is ferruginous, and the crown alone yellowiſh; in ſome it is duſky ferruginous, and the crown and throat yellowiſh¹; in a few inſtances the whole bird is of an uniform duſky ferruginous. The egg is whitiſh with dirty ſpots, ſprinkled with ſome duſky ſpots.

a ſmall

THE MOOR BUZZARD.

a fmall height above the furface of the ground in the bufhes, or even in hillocks covered with thick herbage. It lays three eggs, fometimes four; and though it appears to be more prolific than the Common Buzzard, and, like it, is a ftationary bird, a native of France, where it continues the whole year, it is however more rare, or at leaft more difficult to be found.

Though the Moor Buzzard refembles the Black Kite in many refpects, we muft take care not to confound them; for the Moor Buzzard has, like the Common Buzzard, the Honey Buzzard, &c. a fhort thick neck; whereas the Kite has a much larger one. And the Moor Buzzard is diftinguifhed from the Common Buzzard: firft, by the places it haunts; fecondly, by its flight, which is more rapid and fteady; thirdly, becaufe it never perches upon large trees, but refts upon the ground, or in the bufhes; fourthly, by the length of its legs, which, like thofe of the Hen-harrier and Ringtail, are proportionally taller and flenderer than thofe of the other birds of rapine.

The Moor Buzzard prefers water-fowl; as divers, ducks, &c. It catches fifh alive, and tranfports them in its talons; and when nobler prey fails, it feeds upon reptiles, toads, frogs, and aquatic infects. Though fmaller than the Common Buzzard, it procures a more plentiful fubfiftence; probably becaufe it is more active and vigorous in its movements, and has a

keener

keener appetite : it is alfo more courageous.
Belon afferts, that he has feen it trained to hunt
and catch rabbits, partridges, and quails. It
flies more heavily than the Kites ; and, when it
is purfued by the Falcons, it does not rife into
the air, but flies in a horizontal direction. A
fingle Falcon is not able to get the better of it,
and it would require two or three to be let loofe
at once ; for, like the Kite, it meets its antagonift, but makes a more vigorous and bold defence. The hobbies and the keftrils are afraid
of it, decline the conflict, and even fly its approach.

FOREIGN BIRDS,

WHICH RESEMBLE THE KITE, THE BUZZARDS, AND THE RINGTAIL.

I.

THE bird which is named by Catefby the *Swallow-tailed Hawk**, and by Briffon the *Carolina Kite*. This bird, fays Catefby, weighs fourteen pounds; its bill is black and hooked; but it has no whifkers on the fides of the upper mandible, as the other Hawks; its eyes are very large and black, and the iris red; the head, the neck, the breaft, and the belly, are white; the fhoulders and the back are of a deep purple, but more brownifh below, with a green tint; the wings are long in proportion to the body, and when expanded, meafure four feet; the tail of a deep purple, mixed with green, and much forked, the longeft feather on the fides exceeding by eight inches the fhorteft of the middle. Thefe birds continue long on wing,

---

* *Falco Furcatus*, Linn. *Falco Peruvianus Cauda Furcata*, Klein. *Hirundo Maxima Peruviana*, Feuil.

Its fpecific character :—" Its cere dull coloured; feet yellowifh; " the body dufky above, and whitifh below; the tail forked, and " very long." It is fmaller than the Kite, but as long. The irides are red; the head and neck fnowy; the back dufky or black, fhining with purple or green.

like

like the swallows, and in their flight catch
beetles, flies, and other insects on the trees and
bushes. It is said that they prey upon lizards
and serpents, which have induced some to call
them *Snake-Hawks*. I believe, subjoins Catesby,
that they are birds of passage, never having seen
them during the winter.

We shall only observe, that the bird here
mentioned is really not a Hawk, having neither
the shape nor the instincts. In both these characters it bears a much closer resemblance to the
Kite; and, if we must not consider it as a variety of the European sort, we may at least conclude that it is far more allied to that bird than
to the Hawk.

II.

The bird called by the Indians of Brazil *Caracara*, and of which Marcgrave gives a figure
and a very short description; for he contents
himself with saying, that the *Caracara* of Brazil,
named *Gavion* by the Portuguese, is a kind of
Sparrow-hawk, or small Eagle *(Nisus)*, of the
size of a Kite; that its tail measures nine inches,
its wings fourteen, and reach not so far as the
end of the tail; the plumage rusty, and spotted
with white and yellow points; the tail variegated with white and brown; the head similar
to that of the Sparrow-hawk; the bill black,
hooked, and moderately large; the feet yellow,
the claws like those of the Sparrow-hawk, with
nails that are semilunar, long, black, and very
sharp;

sharp; the eyes of a fine yellow. He adds, that this bird is very destructive to poultry, and that it admits of a considerable variety, some individuals having the breast and belly white.

III.

The bird found in the tracts contiguous to Hudson's-bay, which Edwards terms the *Ash-coloured Buzzard*, and which he describes nearly in the following words : — This bird is of the bulk of a cock, or a middle-sized hen; it resembles the Common Buzzard in its shape and the disposition of its colours; the bill and the cere are of a bluish leaden-colour; the head and the upper part of the neck are covered with white feathers, spotted with deep brown in their middle; the breast is white, like the head, but it is mottled with larger brown spots; the belly and sides are covered with brown feathers, marked with white round or oval spots; the legs are clothed with soft white feathers, speckled irregularly with brown; the coverts of the under part of the tail are radiated transversely with black and white; all the upper parts of the neck, of the back, of the wings, and of the tail, are covered with feathers of a brown cinereous colour, deeper in the middle, and lighter near the edges; the coverts of the lower parts of the wings are of a dark brown, with white spots; the feathers of the tail are barred above with narrow lines of an obscure colour, and barred below with white lines; the legs and feet are cinereous bluish; the nails are black, and the legs covered half their length with feathers of a dull colour. Edwards adds, that this bird, which is found about Hudson's-bay, preys chiefly upon the white grous. After comparing this bird as thus described with the Common Buzzards, the Ringtails, the Harpies, and the Moor-Buzzards, it appeared to us to differ from them all, by the shape of its body, and the shortness of its legs. It has the port of the Eagle; its legs are short like those of the Falcon, and blue like those of the Lanner. We ought therefore to refer it to the genus of the Falcon or of the Lanner, rather than to that of the Buzzard. But as Edwards is one of the ablest ornithologists, we have relinquished our opinion and adopted his; and for this reason we have placed this bird after the Buzzards.

THE SPARROW-HAWK.

The SPARROW-HAWK*.

L'Epervier, Buff.
Falco Nifus, Linn.
Accipiter, Briff.
Accipiter Fringillarius, Ray, Will. and Klein.
Nifus Striatus, Sagittatus, Frif.
Lo Sparviero, Cett.
Sperver, Gunth.

THOUGH nomenclators have reckoned feveral fpecies of Sparrow-hawks, we are of opinion that they may all be reduced to one. Briffon mentions four fpecies, or varieties; viz. the Common Sparrow-hawk, the Spotted Sparrow-hawk, the Small Sparrow-hawk, and the Lark Sparrow-hawk. But we have difcovered that this Lark Sparrow-hawk is only a female Keftril. We have alfo found that the Small Sparrow-hawk is but the Tiercel, or male of the Common Sparrow-hawk; fo that there remains only the Spotted Sparrow-hawk, which is merely an accidental variety of the common fpecies of the Sparrow-hawk. Klein is the firft who has mentioned this variety; and he fays, that it was fent to him from the country of Marienbourg: we ought therefore to refer the Small Sparrow-

* The Greek epithet is Σπιζιας, *Fringillarius*; and the Latin appellation, *Accipiter Fringillarius*, becaufe it preys upon chaffinches *(fringillæ)* and other fmall birds.

hawk

hawk and the Spotted Sparrow-hawk to the
common fpecies, and exclude the Lark Sparrow-hawk, which is only the female Keftril.

It appears that the Sorrel Tiercel of the Sparrow-hawk, N° 466, pl. Eul. differs from the
Haggard Tiercel, N° 467, pl. Eul. the breaft
and belly of the former being much whiter, and
with a much fmaller mixture of ruft-colour than
in the latter, in which thefe parts are almoft entirely ruft, and croffed with brown bars ; in the
former the breaft is marked with fpots, or with
much more irregular bars. The male Sparrow-
hawk is called *Taffel** by the falconers ; its back
receives more brown as it grows older, and the
tranfverfe bars on the breaft are not very regular
till it has undergone the firft or fecond moult :
the fame may be obferved of the female, N° 412,
pl. Eul. To give a fuller idea of the changes in
the diftribution of the colours, we fhall remark
that the fpots on the breaft and belly of the Sorrel Tiercel are almoft all detached, and form rather the figure of a heart, or rounded triangle,
than a continued and uniform fucceffion of a
brown colour, fuch as we perceive in the bars
on the breaft and belly of the Haggard Tiercel,
that is of the Tiercel which has had two moultings : the fame changes happen in the female,
in which the brown tranfverfe belts are in the
firft year only unconnected fpots. It will be
found in the following article that the Gos-

* *Mouchet.*

hawk

hawk is ftill more remarkable for the variations of colour. Nothing more clearly demonftrates that the characters which our nomenclators have drawn from the diftribution of colours are in-fufficient, than that the fame bird has the firft year fpots or brown longitudinal bars extend-ing downwards; and the fecond year is marked with tranfverfe belts of the fame colour. This fingular change is more ftriking in the Gos-hawk, and in the Sparrow-hawks; but it occurs in a certain degree alfo in other fpecies of birds. In fhort, all the fyftems that are founded upon difference of colour and diftribution of fpots, are in the prefent cafe entirely futile.

The Sparrow-hawk continues the whole year in our provinces. The fpecies is numerous; I have received many in the depth of winter that had been killed in the woods; at that time they were very lean, and weighed only fix ounces. They are nearly of the fize of a magpie. The female is much larger than the male; fhe builds her neft on the loftieft trees of the foreft, and generally lays four or five eggs, which are fpot-ted with a yellow reddifh near the ends. The Sparrow-hawk is docile, and can be eafily train-ed to hunt partridges and quails; it alfo catches pigeons that ftray from their flock, and commits prodigious devaftation on the chaffinches, and other fmall birds which troop together in win-ter. It is probable that the Sparrow-hawk is more numerous than we fuppofe; for befides

N 3   thofe

thofe that remain the whole year in our climate, it appears at certain feafons to migrate in immenfe bodies to other countries *; and the fpecies is found fcattered in the ancient continent †, from Sweden ‡ to the Cape of Good Hope ‖.

* I muft here tranfcribe a paffage of confiderable length from Belon, which proves the migration of thefe birds, and even points out the time when they begin their flight:

" We were at the mouth of the Pontus Euxinus, where begins the ftrait of the Propontis: we afcended the higheft mountain, and found a bird-catcher very fuccefsfully employed; and, as it was about the end of April, when no birds can build their nefts, we thought it ftrange that fo many Kites and Sparrow-hawks fhould flock thither. The bird-catcher was very affiduous, and fcarcely allowed one to efcape; he caught more than a dozen in an hour. He was concealed behind a bufh, and in front, about two or three paces diftant, he had conftructed a clofe fquare airy meafuring two paces; round it were fixed fix ftaffs, three on each fide, an inch thick, and about a man's height, and on the top of each a notch was cut; a very fine green net was faftened to thefe notches, and fpread a man's height from the ground; in the middle of the fpace was a ftake of a cubit high, and to the top of which was attached a cord that led to the perfon concealed behind the bufh; to this cord feveral birds were faftened, and fed on grain within the airy. When the bird-catcher perceived the Sparrow-hawk advancing from the fea, he fcared thefe birds; and the Sparrow-hawk, whofe fight is fo keen as to defcry them at the diftance of half a league, fhot with expanded wings to feize his prey, and in the eagernefs and rapidity of his motion was entangled in the net. The perfon then took the bird, and flipped it into a linen bandage ready fewed, which confined the wings, thighs, and the tail, and threw it upon the ground, where it could not ftir. We could not conceive whence the Sparrow-hawks came, for during the two hours that we ftaid, more than thirty were caught; fo that in a day one man might catch above a hundred. The Kites and Sparrow-hawks arrived in a chain that extended as far as the eye could reach." BELON, *Hift. Nat. des Oifeaux.*

† The Sparrow-hawks are common in Japan, as well as in every part of the Eaft Indies. KÆMPFER.

‡ LINNÆUS, *Fauna Suecica.*

‖ Kolben.

[A] The specific character of the Sparrow-hawk, *Falco Nisus*, LINN. is " That its cere is greenish ; its feet yellow ; the abdo-" men white, waved with grey, and blackish streaks on the tail." There are two varieties belonging to it.

First, The Spotted Sparrow-hawk. Its back is earthy-coloured, sprinkled with white spots ; the under part of the body is squamous, and more deeply stained ; the under surface of the wings and of the tail is varied with broad white zones, and dirty narrow stripes.

Secondly, The White Sparrow-hawk, which has been killed in England.

The GOS-HAWK.

L'Autour, Buff.
Falco Palumbarius, Linn. Gmel. Will. Klein, &c.
Aftur, Briff.
Groffe Gepfeilter Falck, Frifch.
In Italian, *Aftore*.
In German, *Habicht, Groffer Habicht*.
In Polifh, *Jaftrzabwielki* *.

THE Gos-hawk is a beautiful bird, much
larger than the Sparrow-hawk, which it
however refembles by its inftincts, and by a
common character, that, in the birds of rapine,
is confined to them and the Butcher-birds: this
is, that their wings are fo fhort as not to reach
near the end of the tail. It refembles the Sparrow-hawk alfo by another circumftance;—the
firft feather of the fhort wing is rounded at the
tip, and the fourth feather of the wing is the
longeft of all. Falconers divide thefe birds of
fport into two claffes; viz. thofe of falconry,
properly fo called, and thofe of *hawking* †: and
in this fecond clafs they include not only the
Gos-hawk, but the Sparrow-hawk, the Harpies, the Buzzards, &c.

* The Greek epithet is, Αϛερίας, *Stellaris* ; and the Latin appellation, *Accipiter Stellaris*.

† *De l'autourferie.*

The

THE GOSHAWK.

The Gos-hawk, before it has fhed its feathers, that is, in its firft year, is marked on the breaft and belly with longitudinal brown fpots ranged vertically; but after it has had two moultings, thefe difappear, and their place is occupied by tranfverfe bars, which continue during the reft of its life. Hence we are apt to be deceived with refpect to this bird, from the change that happens in the difpofition of the colours of the plumage. N° 461, pl. Eul. is a young one; N° 418, an old one.

The Gos-hawk is furnifhed with longer legs than other birds to which it bears a clofe analogy; as the White Jer-Falcon, which is nearly of the fame fize: the male is much fmaller than the female: both are carried on the hand, and not ufed as decoys; they foar not fo high as thofe whofe wings are longer in proportion to their body; they have many habits in common with the Sparrow-hawk, yet they do not dart directly downwards upon their prey, but catch it by a fide fhoot. It appears by Belon's account that the Gos-hawk can be enfnared by a contrivance fimilar to what is practifed againft the Sparrow-hawk. A white pigeon, which can be perceived at a great diftance, is placed between four nets, nine or ten feet high, inclofing a fpace of nine or ten feet each way round the pigeon, which is in the centre, the Gos-hawk defcends obliquely, (a proof that he makes only fide attacks,) pufhes the net to reach his prize, and though

6   entangled,

entangled, he devours it, and till fated makes few attempts to efcape.

The Gos-hawk is found in the mountains of Franche Compté, of Dauphinè, of Bugey, and even in the forefts of the province of Burgundy, and in the neighbourhood of Paris; but it is ftill more common in Germany than in France, and the fpecies feems to penetrate in the countries of the north as far as Sweden, and advance in thofe of the eaft and fouth, to Perfia and Barbary. Thofe bred in Greece are, according to Belon, the beft of all for falconry. " They have," fays he, " a large head, a thick neck, and much " plumage. Thofe of Armenia," he adds, " have " green eyes ; in thofe of Perfia, they are light- " coloured, hollow and funk ; in thofe of Africa, " which are lefs efteemed, they are at firft black, " and after moulting, become red." But this character is not peculiar to the Gos-hawks of Africa ; thofe of our own climate have eyes which affume a deeper red as they advance in age. There is, in the Gos-hawks of France, a difference or variety even of plumage and colour, which has drawn naturalifts into a fort of miftake. They have applied the name of Moor Buzzard *(Bufard)* to a Gos-hawk, whofe plumage is light-coloured, and which is more indolent than the Brown Gos-hawk, and not fo eafily trained. It is, however, undoubtedly a Gos-hawk, though the falconers reject it. This light-coloured Gos-hawk admits even a flight

variety,

variety, where the wings are fpotted with white, from which circumftance it has been called the *Variegated Moor Buzzard*. But both thefe birds are really Gos-hawks.

I kept for a long time a male and a female of the Brown Gos-hawk: the female was at leaft a third larger than the male, and its wings, when clofed, did not reach within fix inches of the end of the tail: it was more bulky at four months old, which I conceive to be the term of the growth of thefe birds, than a large capon. During the firft five or fix weeks, thefe birds were of a grey white; the back, the neck, and the wings, became gradually brown; the belly and the under-part of the throat did not change fo much, and were generally white or yellowifh white, with longitudinal brown fpots the firft year, and tranfverfe brown bars the following years. The bill is of a dirty blue, and the cere is of a leaden colour; the legs are featherlefs, and the toes of a deep yellow; the nails are blackifh, and the feathers of the tail, which are brown, are marked with very broad bars of a dull grey colour. During the firft year, the feathers under the throat are in the male mottled with a reddifh colour, by which circumftance it differs from the female; though, if we except the fize, it clofely refembles it in other re-fpects.

It was obferved, that though the male was much fmaller than the female, it was fiercer and

more

more vicious; they were both difficult to tame; they fought often, but rather with their claws than with their bill, which they seldom employ but to tear the birds or other small animals that they want to catch; they turn upon their back and defend themselves with their spread talons. Though confined in the same cage, they were never perceived to contract the least affection for each other. They continued together a whole summer, from the beginning of May to the end of November, when the female in a violent fit of rage, murdered her mate, at nine or ten o'clock in the evening, when the silence of night had soothed the rest of the feathered race in profound repose. Their dispositions are so bloody, that if a Gos-hawk be left with several Falcons, it butchers them all, one after another. It appears, however, to prefer the common and field mice and small birds, and eagerly devours raw flesh, but constantly declines meat that has been cooked; however, by long fasting, it can be brought to overcome this natural aversion. It plucks the birds very neatly, and tears them into pieces before it feeds; but it swallows the mice entire. Its excrements are whitish and watery; it often disgorges the skins of the mice rolled together. Its cry is raucous, ending always in sharp notes, the more disagreeable the oftener they are repeated; it discovers a constant uneasiness when a person approaches; it startles at every thing; so that a

person

perſon cannot paſs near the cage where it is kept, without throwing it into violent agitations, and occaſioning repeated ſcreams. [A]

[A] The ſpecific character of the Gos-hawk *(Falco Palum-barius,* LINN.) is, "That its cere is black, the margin and " feet yellow, the body duſky, the tail feathers marked with pale " bars, the eye-brows white." It is larger than the Common Buzzard, being one foot ten inches long, but it is of a ſlender and more elegant ſhape.   It is found in Europe, Aſia, and America.

FOREIGN BIRDS,

THAT ARE RELATED TO THE SPARROW-HAWK AND GOS-HAWK.

I.

THE bird which we have received from Cayenne without any name, and which we have termed *Thick-billed Sparrow-hawk of Cayenne**, (N° 464, pl. Eul.) for it refembles the Sparrow-hawk more than any other bird of prey; being only fomewhat larger and rounder fhaped. Its bill is alfo thicker and longer, but the legs rather fhorter. The lower part of the throat is of an uniform wine colour; whereas, in the Sparrow-hawk it is white, or whitifh: but in general the refemblance is fo clofe, that we may confider it as a kindred fpecies, and perhaps the difference originates from the influence of climate.

II.

The bird fent from Cayenne without a name, and to which we have given that of the *Little Gos-Hawk of Cayenne*, becaufe it was confidered by fkilful falconers as of the Gos-hawk kind. I muft indeed own, that it appeared to us to have more refemblance to the Lanner, as de-

* *Epervier à gros bec de Cayenne.*

fcribed

fcribed by Belon, than to the Gos-hawk; for its legs are fhort and of a blue colour, which are two characters of the Lanner: but perhaps it is neither the one nor the other. We every day commit miftakes in attempting to refer the birds and quadrupeds of foreign countries to thofe of our own climate: and fuch may be the cafe in the prefent inftance.

III.

The Carolina bird, defcribed by Catefby under the name of *Pigeon-hawk*, which is more flender than the common Sparrow-hawk. Its iris, cere, and feet yellow; its bill whitifh at its origin, and blackifh near the hook; the upper part of the head, neck, back, rump, wings, and tail covered with white feathers, mixed with fome brown ones; the legs clothed with long white plumage, tinged flightly with red, and variegated with longitudinal brown fpots. The feathers of the tail are brown like thofe of the wings, but marked with four white tranfverfe bars.

The JER-FALCON.

Le Gerfaut, Buff.

THE Jer-falcon, both in its figure and its
dispositions, deserves to be ranked the first
of all the birds of falconry. It exceeds them all
in point of size, being at least equal in bulk to the
Gos-hawk. It differs from them by certain
general and invariable characters which belong
exclusively to those peculiarly calculated for
sport. These noble birds are, the Jer-falcons,
the Falcons, the Sacres, the Lanners, the Hobbies, the Merlins, and the Kestrils; their wings
are almost as long as their tail; the first feather
of the wing, called *the * hoop,* is nearly as long
as that inserted next it, and about an inch of
the extremity is sharpened into a figure resembling the blade of a knife. In the Gos-
hawks, the Sparrow-hawks, the Kites, and the
Buzzards, the tail is longer than the wings, and
the first feather of the wing is much shorter,
and is rounded at the end. Besides, the longest
feather in these is the fourth of the wing, but
it is the second in the former. We may add,
that the Jer-falcon differs from the Gos-hawk

---

* *Cerceau.*

also

THE WHITE JERFALCON .

also by its bill and feet, which are bluiſh, and
by its plumage, which is brown on all the upper
part of the body, and white ſpotted with brown
on the under, and its tail, which is gray, and
barred with duſky lines. (N° 210, Pl. Enl.) This
bird is common in Iceland, and it appears that
there is a variety in the ſpecies; for we have
received from Norway a Jer-falcon, which is
found in all the arctic regions, (N° 462, Pl. Enl.)
but differs ſomewhat from the other by the
ſhades and diſtribution of the colours, and which
is more eſteemed by the falconers than that of
Iceland, on account of its greater courage, activity, and docility.   But there is another variety, (Pl. Enl. N° 446.) which is entirely white,
and which, if it were not found alike in all the
dreary tracts of the north, might be aſcribed to
the influence of the climate.   Intelligent falconers inform me, that the young have the ſame
colour, which they always retain; we can
neither attribute the change therefore to extreme
age nor exceſſive cold.   It is therefore probable,
that there are three diſtinct and permanent breeds
of the Jer-falcons; viz. the Iceland Jer-falcon,
the Norwegian Jer-falcon, and the White Jer-
falcon.   Theſe birds are natives of the inhoſ-
pitable arctic regions, both in Europe and in
Aſia; they inhabit Ruſſia, Norway, Iceland, and
Tartary, but are never found in the warm or
even temperate countries.   Next to the Eagle
it is the moſt formidable, the moſt active, and the

most intrepid of all the rapacious birds; and it is also the dearest and the most esteemed for falconry. It is transported from Iceland and Russia into France *, Italy, and even into Persia and Turkey †; nor does the heat of these climates appear to diminish its strength or blunt its vivacity. It boldly attacks the largest of the feathered race; the stork, the heron, and the crane, are easy victims: it kills hares by darting directly down upon them.—The female, as in the other birds of prey, is much larger and stronger than the male, which is called the *Tiercel Jer-Falcon*, and is used in falconry only to catch the kite, the heron and the crows. [A]

* We should not see the Jer-falcon were it not brought from a foreign country; it is said to come from Russia, where it breeds, and does not inhabit France or Italy, and is a bird of passage in Germany.——It may be flown against any thing, and is bolder than any other bird of prey. BELON.

† The following passage seems to refer to the Jer-falcon: We must not omit to mention a bird of prey which comes from Muscovy, whence it is transported into Persia, and which is almost as large as an Eagle. These birds are rare, and only the king is permitted to keep them. As it is customary in Persia to estimate all the presents without exception that are made to the king, these birds are rated at one hundred *tomans* a-piece, which answers to one thousand five hundred crowns; and if any of them die on the road, the ambassador brings the head and the wings to his majesty. It is said that this bird makes its nest in the snow, which it melts to the ground by the heat of its body, sometimes to the depth of a fathom, &c. CHARDIN.

[A] Linnæus makes two species of the Jer-falcon: the first is *Falco Gyrfalco*, or Brown Jer-falcon; and is perhaps Buffon's
Norwegian

Norwegian Jer-falcon: " Its cere is cœrulean, its feet yellowish, its body dusky, with cinereous stripes below, and the sides of the tail white." The second species is the *Falco Candidus*, or White Jer-falcon, and seems to be the same with that of Buffon. Its character is, " That its cere and feet are of a cœrulean cast, verging to cinereous, its body white, with dusky spots." And to this belongs a variety, which is the Iceland Jer-falcon, in which the body is dusky, with white spots on the back and wings, and below it is white spotted with black, and the feet are yellow.

The LANNER.

Le Lanier, Buff.
Falco Laniarius*, Linn. Gmel. Ray, Briff. Klein, &c.
Lahneret, Alb.
In German, *Swimer*, or *Schmeymer*.
In Italian, *Laniero*.

THIS bird, which Aldrovandus calls *Laniarius Gallorum*, and which Belon fays is a native of France, and more ufed by the falconers than any other, is now become fo rare, that we could not procure a fpecimen of it. It occurs in none of our cabinets, nor is it found in the feries of coloured birds by Edwards, Frifch, and the authors of the Britifh Zoology. Belon himfelf, though he defcribes it at confiderable length, does not give the figure; and it is the fame with Gefner, Aldrovandus, and the other modern naturalifts.—Briffon and Salerne confefs that they never faw it; and the only figure that we have of it is in Albin, whofe plates are known to be wretchedly executed. It appears then, that the *Lanner*, which is now fo rare in France, has always been fo in Germany, England, Switzerland, and Italy, fince

* The name *Laniarius*, or *Lanner*, is derived from *laniare, to tear*; becaufe the bird mangles cruelly the poultry and other victims of its rapine.

the

the authors of thefe countries mention it upon
the authority of Belon. It is however found in
Sweden, for Linnæus ranges it among the native birds of that country ; but he gives only a
flight defcription, and totally omits its hiftory.
All the information that we can obtain is from
Belon, and we fhall therefore tranfcribe his account. " The Lanner, or Lanner-falcon," fays
he, " generally conftructs its aerie, in France, on
" the talleft trees of the forefts, or on the moft
" elevated rocks. As its difpofitions are more
" gentle and its habits more flexible than the
" Common Falcons, it is ufed for *every pur-*
" *pofe.* It is lefs corpulent than the Genteel
" Falcon, and its plumage is more beautiful
" than that of the Sacre, efpecially after moult-
" ing ; it is alfo fhorter than the other Falcons.
" The falconers prefer the Lanner that has a
" large head and blue-*bordered* feet; it flies
" both on rivers and on the plains. It fubfifts
" better than any other Falcon upon coarfe flefh.
" It is eafily diftinguifhed, for its bill and feet
" are blue; the feathers on the front mottled
" with black and white, with fpots ftretching
" along the feathers, and not tranfverfe as in
" the Falcon. . . . When it fpreads its wings,
" the fpots feen from below appear different
" from thofe of the other birds of prey ; for
" they are fcattered and round *like fmall pieces of*
" *money (deniers).* Its neck is fhort and thick;
" as alfo its bill. The female is called *Lanner*,

O 3   " and

" and is much larger than the male, which is
" named *Lanneret*; they are both similar in the
" colour of their plumage. It remains in the
" country the whole year, and no bird is so
" faithful to its favourite haunt. It is easily
" trained to catch the Crane: the best time for
" sport is after moulting, from the middle of
" July to the end of October; but the winter
" is an improper season." [A]

[A] The specific character of the Lanner, *Falco Laniarius*,
Linn. is, that " its cere is yellowish, its feet and bill cœrulean,
its body marked beneath with black longitudinal spots." It inhabits Europe, but is not common in England; it is frequent in
Iceland, the Feroe islands and Sweden, in the Uralian chain and
other parts of Tartary, though not found in the east and north of
Siberia. It breeds in low trees: it is smaller than the Buzzard.

THE SACRE RFALCON.

The S A C R E.

Le Sacre, Buff.
Falco Sacer, Gmel. Briff. Will. Klein, &c.
In German, *Sacker.*
In Italian, *Sacro.*

I HAVE removed this bird from the Falcons,
and placed it after the Lanner ; though some
of our nomenclators confider it only as a variety
of the fpecies of Falcons : becaufe, if we reckon
it a mere variety, we ought to refer it to the Lanners rather than to the Falcons.   Like the Lanner, the feet and bill of the Sacre are blue ;
while thofe of the Falcons are yellow.   This
character, which appears fpecific, would incline
us to conclude that the Sacre is but a variety of
the Lanner ; but they differ widely in their fize
and the colour of their plumage, and feem rather
to be two diftinct though proximate kinds.   It
is fomewhat fingular that Belon is the only one
who has noticed the diftinguifhing marks of this
bird, and, without his affiftance, naturalifts
would be fcarcely, if at all, acquainted with the
Sacre and the Lanner.   Both have become very
rare, and it is probable that their inftincts are
the fame, and confequently that they are kindred
tribes.   But as Belon examined thefe birds, and
has defcribed them really diftinct, we fhall

O 4   tranfcribe

tranfcribe his account of the Sacre, as we have
already done that of the Lanner :—The plumage
of the Sacre is inferior in beauty to that of the
other birds of falconry; being of a dirty ferruginous colour, like that of the Kite. It is
low, its legs and toes blue, in fome degree fimilar to the Lanner. It would be equal to the
Falcon in bulk, were it not more compact and
rounder fhaped. It is a bird of intrepid courage,
and comparable in ftrength to the Pilgrim Falcon;
it is alfo a bird of paffage, and it is rare to find
a man who can boaft that he has ever feen the
place where it breeds. Some falconers are of
opinion that it is a native of Tartary and Ruffia,
and towards the Cafpian Sea; that it migrates towards the fouth, where it lives part of
the year; and that it is caught by the falconers
who watch its paffage in the iflands of the Archipelago, Rhodes, Cyprus, &c. And as the
Sacre can be made to foar after the Kite, it can
alfo be trained for rural fport, to catch wild
geefe, *buftards*, *olives*, pheafants, partridges, and
every fort of game. The *Sacret* is the male, and
the *Sacre* the female; the only difference between
which confifts in fize.

If we compare this defcription of the Sacre
with what the fame author has given of the
Lanner, we fhall eafily perceive, firft, that thefe
two birds are nearer related to each other than
to any other fpecies: fecondly, that they are
birds of paffage; though Belon fays that, in his
                                                    time,

THE YOUNG FALCON.

being juſt. Their inſtinctive diſpoſitions have not, in the leaſt, been altered by man; and, though ſubſervient to his pleaſures, and flattering to his vanity, they ſtill retain the native ſenſe of independence, and refuſe to multiply and tranſmit poſterity under his dominion. The original ferocity of theſe birds is indeed broken by careful attention and multiplied reſtraints. They are obliged to purchaſe their exiſtence by performing taſks that are exacted; not a morſel of food is granted but for a ſervice received; they are fixed, pinioned, muffled; they are even excluded from light, and totally denied ſubſiſtence, to render them more dependent, more docile, and to add to their natural vivacity the urgency of want. But they ſerve from neceſſity and from habit, not from attachment; they remain captives, but never become domeſtics; the individual alone feels the weight of ſlavery; the ſpecies preſerves its liberty conſtantly untainted, and never owns the empire of man. It requires the moſt watchful attention to ſurpriſe ſome ſtraggling priſoners; and nothing is more difficult than to ſtudy their œconomy in the ſtate of nature. As they inhabit the moſt rugged precipices on the loftieſt mountains, and ſeldom alight upon the ground, but ſoar in the aerial regions, and fly with unequalled rapidity, few facts can be diſcovered with reſpect to their primitive inſtincts. It has only been obſerved, that they prefer breeding in rocks of a ſouthern

expo-

expofure; that they build their nefts in the moft inacceffible *holes* and *caverns*; that they commonly lay four eggs in the latter months of the winter, and fit but a fhort time; for the young are adult about the fifteenth of May, and change their colour according to their fex, their age, and the feafon of moulting; that the females are much larger than the males; that the parents utter piercing, difagreeable, and almoft inceffant fcreams, when they expel their young; to which violent remedy they have recourfe, like the Eagle, from hard neceffity, which breaks the bonds of families, and diffolves the union of every fociety, as foon as the tracts inhabited afford not a fufficient fubfiftence.

The Falcon is perhaps that bird whofe courage, compared with its ftrength, is the moft open and the moft confpicuous. It darts directly downwards without deviating from the perpendicular; while the Vulture, and moft of the other birds of rapine, furprife their prey by an oblique defcent. It alights vertically upon the feathered victim entangled in nets, kills it, and devours it upon the fpot; or, if not too large, it carries the carcafs aloft into the air. It prefers pheafants for its prey; and if it difcovers a flock of them, it drops fuddenly among them as if it fell from the clouds, becaufe it defcends from fuch an amazing height in fo fhort a time that its vifit is always unexpected. It frequently attacks the Kite, either to amufe its courage, or to

feize

THE HAGGARD FALCON.

seize its prey; but this is rather a contemptuous
insult than an obstinate combat. It treats its
enemy as a coward, pursues it, strikes it with
disdain, and as it meets with but feeble resistance, it allows the Kite to escape with its life,
being as much disgusted perhaps with the rank-
ness of its carcass, as conciliated by the mean-
ness of its conduct.

Those who inhabit the vicinity of our great
mountains in Dauphiny, Bugey, and Auvergne,
and the foot of the Alps, can ascertain the veracity of all these facts. There have been sent to
the King's falconry from Geneva young Falcons
that had been caught in the adjoining mountains
in April, and which appeared to have acquired
their full size and vigour before the month of
June. When they are young they are called
Sorrel Falcons, because they are then browner
than in the following year (Pl. Enl. N° 470.);
and the old Falcons, which are much whiter
than the young, are termed *Haggards* (Pl. Enl.
N° 421.). The Falcon represented in the last
plate appears to be hardly two years old, and
has still a great number of brown spots on the
breast and belly; for in the third year these
spots diminish, and the quantity of white on
the plumage increases (Pl. Enl. N° 430.).

As these birds every where seek the highest
rocks, and as most islands are but groups and
points of mountains, they abound in Rhodes,

8   Cyprus,

Cyprus, Malta, and in the other iflands of the Mediterranean, and even in the Orkneys and Iceland : but, according to the different climates which they inhabit, they admit of varieties, which it will be proper to mention.

The Falcon which is a native in France is about the fize of a hen ; its extreme length is eighteen inches ; its tail is five inches ; and its wings when fpread are three feet and a half, and when clofed reach to the end of the tail. It is unneceffary to take notice of the colours of the plumage, becaufe they vary with the age. I fhall only remark, that the feet are commonly green, and that when the feet and the cere are yellow, they receive the name of *Yellow-billed-Falcons* (Pl. Enl. N° 430.), and are confidered as inferior to the others, and deemed unfit for the fport. The Tiercel is employed to catch partridges, magpies, jays, blackbirds, and others of that kind ; but the female is engaged in the nobler chace of the hare, the kite, the crane, and other large birds.

It appears that this fpecies of Falcon, which is very common in France, is found alfo in Germany. Frifch has given a coloured figure of the Sorrel Falcon with yellow feet and cere, by the name of *Enter-floffer*, or *Schwartz-braune Habicht* (i. e. *Plunderer of ducks*, or *Black-brown Hawk*) ; but he is miftaken in terming it *Brown Gos-hawk (Autour)* ; for it differs from that bird

by

by its fize and inftincts *.   It feems that thefe
occur alfo in Germany, and fometimes in France;
another fpecies, which is the Rough-footed Falcon with a white head, and which Frifch improperly calls *Vulture*.   " This Vulture is completely
" clothed with feathers on the feet, in which cir-
" cumftance it differs from all the diurnal rapa-
" cious birds that have a hooked bill. The Rock-
" eagle is furnifhed with fimilar feathers, but they
" only reach half-way to the feet. The nocturnal
" birds of prey, fuch as the owls, are indeed co-
" vered to the nails ; yet this is rather a fort of
" down. This *Vulture* chafes every kind of prey,
" though it never grovels among dead carcaffes."
It feeds not upon carrion, becaufe it is not a
Vulture but a Falcon ; and fome of our naturalifts have confidered it as only a variety of the
common fpecies in France.   It bears indeed a
clofe refemblance, and differs chiefly by the
whitenefs of its head ; but the character, that its
feet are covered with feathers to the nails, feems
to be fpecific, or at leaft to indicate a conftant
independent variety.

A fecond variety is the White Falcon, which
occurs in Ruffia, and perhaps in other countries
of the north.   Some of this fort are of an uniform white, except at the ends of the great fea-

* This remark of Buffon's feems to be groundlefs. The German word *Habicht* is generic, and fignifies any kind of Hawk. It
is probably the fame with the Saxon *Hafoc*, from which the Englifh
term *Hawk* is derived. The Welfh appellation *Hæbeg* is ftill more
analogous in regard to found, T.

thers of the wings, which are blackiſh; others
are alſo entirely white, except a few brown ſpots
on the back and wings, and a few brown ſtripes
on the tail *. Since this Falcon differs from the
common kind in nothing but the whiteneſs of
its plumage, we may conſider it as merely a variety occaſioned by the general influence of extreme cold. Yet in Iceland there are Falcons
which have the ſame colour with ours, and are
only ſomewhat larger, and have their wings and
tail longer: theſe then ought not to be ſeparated
from the common ſpecies. The ſame remark
may be made in regard to the *Genteel Falcon*,
which moſt naturaliſts have ſtated as different;
in faʄt, the epithet *Genteel* † is applied when the
bird is high bred, and of an elegant ſhape.
Accordingly the old writers on falconry reckoned only two kinds of Falcons; the Genteel Falcon, which is bred in our climates, and the
Pilgrim or Peregrine Falcon, which is of foreign
extraʄtion; and they regarded all the others as
varieties of either of theſe. Some Falcons indeed from foreign countries pay us tranſient vi-
ſits; they appear moſtly on the ſouthern ſhores,
and are caught at Malta; they are for that reaſon called the *Paſſenger Falcons*, and are much
blacker than the common kind. It would appear that this Black Falcon enters into Germany
as well as France; for it is the ſame with the
Brown Falcon of Friſch. It even penetrates to

* Briſſon.   † *Gentle*, in old Engliſh.

more

more diftant climates; and Edwards has figured and defcribed it under the name of the *Black Falcon of Hudfon's-bay.*

To the fame fpecies we may alfo refer the Falcon of Tunis or Carthage, mentioned by Belon, " which," he fays, " is rather fmaller " than the Pilgrim Falcon, its head thicker and " rounder, and in its bulk and plumage like the " Lanner." The Tartary Falcon ought perhaps to have the fame arrangement; which is on the contrary fomewhat larger than the Pilgrim Falcon, and is reprefented by Belon to differ in another circumftance, that the upper part of its wings is rufty, and its toes longer.

To give a condenfed view of the facts which we have confidered in detail: 1. There is in France only one fpecies of Falcon well known, and which conftructs its aerie in the mountain-ous provinces: the fame is found alfo in Ger-many, Poland, Sweden, and as far as Iceland, to the north; and in Italy *, Spain, the iflands in the Mediterranean, and perhaps Egypt †, to the fouth. 2. The White Falcon is merely a variety of the fame fpecies, produced by the influence of a northern climate. 3. The Genteel Falcon is of the fame fpecies with the common kind ‡. 4. The Pilgrim or Paffenger Falcon is

of

---

* Aldrovandus. † Profper Alpinus.

‡ John of Franchieres, who is one of our oldeft writers on fal-conry, and perhaps the beft, reckons only feven fpecies of Falcons; viz. the Genteel Falcon, the Peregrine Falcon, the Tartarian Fal-

of a different species, and perhaps includes some varieties; such as the Barbary Falcon, the Tunis Falcon, &c.

Whatever then the statement of our nomenclators may be, there are only two kinds of Falcons in Europe, the one native and the other to be regarded as foreign. If we survey the numerous catalogue which Brisson has given, we shall find, 1. That his Sorrel Falcon is only the young of the common species: 2. That his Haggard Falcon is an old one of the same: 3. That his Falcon with a white head and rough feet, is a permanent variety of the same kind: 4. That his White Falcon comprehends two, perhaps three, different species of birds; the first and third of which may derive their colour from the general influence of the arctic climates, but the second, which Brisson borrows from Frisch, is undoubtedly not a Falcon, and is only a bird of prey common in France, and named the *Harpy*: 5. That the Black Falcon is the true Pilgrim or Passenger Falcon, which may be regarded as foreign: 6. That the Spotted Falcon

con, the Jer-falcon, the Sacre, the Lanner, and the Tunisian Falcon. If we omit the Jer-falcon, the Sacre, and the Lanner, which are not Falcons, there remain only the Genteel Falcon and the Peregrine Falcon, of which the Falcons of Tartary and Tunis are two varieties. This author knew only one species of Falcon that was a native of France, which he calls the *Genteel Falcon*; and this circumstance confirms what I have before said, that the Genteel and Common Falcons are really the same species.

is

is only the young of the fame: 7. That the
Brown Falcon ought rather to be reckoned a
Moor Buzzard: Frifch is the only one who has
given a figure of it, and he obferves that it fometimes feizes wild pigeons in its flight; that it
foars high; that it is difficult to fhoot, and yet
that it watches the aquatic birds near pools and
marfhes:—thefe circumftances combined would
fhew that it is only a variety of the Moor Buzzards,
though its tail is not fo long as theirs: 8. That
his Red Falcon is only a variety of the Common
Falcon, which, Belon and the old writers on
falconry fay, frequents the fenny tracts: 9. That
his Red Indian Falcon is a foreign bird, of
which we fhall afterwards treat: 10. That his Italian Falcon, the account of which he borrows from
Johnfton, may be regarded as a variety of the
common fpecies of Falcon inhabiting the Alps:
11. That his Iceland Falcon is, as we have already remarked, another variety of the Common
Falcon, and only fomewhat larger: 12. That
the Sacre is not a variety of the Falcon, but a
different fpecies, which muft be treated of feparately: 13. That his Genteel Falcon is really
the Common Falcon, only defcribed at a different
feafon of moulting: 14. That Briffon's Pilgrim
Falcon is the fame, only aged: 15. That the
Barbary Falcon is but a variety of the Foreign
or Paffenger Falcon: 16. And that fo is the
Tartary Falcon: 17. That the Collared Falcon
is a bird of a different genus, which we have

termed

termed *Soubuse* (Ring tail): 18. That the Rock
Falcon is not a Falcon, but is most related to the
Hobby and Kestril, and should therefore be considered apart: 19. That the Mountain Falcon
is only a variety of the Rock Falcon: 20. That
the Cinereous Rock Falcon is only a variety of
the common species of Falcon: 21. That the
Hudson's Bay Falcon is of a different species
from the European: 22. That the Stellated
Falcon is of a different genus: 23. That the
Crested Falcon of India, the Falcon of the Antilles, the Fisher Falcon of the Antilles, and the
Fisher Falcon of Carolina, are all foreign birds,
of which we shall treat in the sequel.

Thus the Falcons are reduced to two species;
the Common or Genteel Falcon, and the Passenger or Pilgrim Falcon.—Let us now con-
sult our old writers on falconry in regard to the
difference of their instincts, and in the proper mode of education. The Genteel Falcon
drops its feathers in March, and even earlier;
the Pilgrim Falcon does not moult until August.
It is broader over the shoulders, its eyes are
larger and deeper sunk, its bill thicker, its legs
longer and better set than in the Genteel Falcon * : those caught in the nest are called *Ninny
Falcons (Faucons-niais)*; when taken too young,
they are often noisy and difficult to train; they
ought not therefore to be disturbed till they are

* Artelouche's Falconry, printed along with the treatises of
Fouilloux, Franchieres, and Tardif. *Paris* 1614.

consider-

confiderably grown ; and if they are to be removed from the neft, they muft not be handled,
but put into another neft as like the original one
as poffible, and fed with bears flefh, which is
common in the mountains where thefe birds are
found, or inftead of that, they may be nourifhed
with the flefh of chickens : without thefe precautions, their wings do not grow *, and their
legs are eafily broken or diflocated. The Sorrel
Falcons, which are the young ones, and which
have been caught in September, October, and
November, are the beft, and the eafieft bred:
thofe which are caught later, in winter or in
the following fpring, and confequently are nine
or ten months old, have tafted too much of
freedom to fubmit patiently to captivity, and
their fidelity or obedience can never be relied on;
they often defert their mafter when he leaft ex-
pects it. The Pilgrim Falcons are caught in
their paffage every year in September on the
iflands in the fea, and the high beaches by the
fhore. They are naturally quick and docile,
and very eafy to train ; they may be flown all
May and June ; for they are late in moulting ;
but when it begins, they are foon ftripped of
their plumage. The Pilgrim Falcons are caught
not only on the coafts of Barbary, but in all the
iflands of the Mediterranean, and particularly

* Catalogue of all the birds of prey employed in falconry, contained in the preceding collection.

that

that of Candia, which formerly furnifhed our beft Falcons.

The art of falconry does not belong to Natural Hiftory; we fhall not therefore enter into details, but refer to the *Encyclopedie* for information on that fubject. " A good Falcon," fays Le Roi, author of the article of *falconry*, " ought to " have a round head, a fhort thick bill, a very " long neck, finewy breaft, broad long thighs, " fhort legs, broad feet, flender tces, that are " lengthened and finewy at the joints, ftrong " incurvated nails, long wings : the marks of " ftrength and courage are the fame in the " Jer-falcon and the Tiercel, which is the male " in all the rapacious birds, and which is fo " called, becaufe it is one third lefs than the " female. A more certain indication of the " goodnefs of the bird, is its riding *(chevaucher)* " againft the wind; that is, briftling againft it, " and fitting firm on the hand when expofed to " it. The plumage of a Falcon fhould be brown " and of an uniform colour : the proper caft of " the fole is fea-green. Thofe whofe fole is " yellow, and whofe plumage is fpotted, are " lefs efteemed; the black ones are prized : but " whatever be their plumage, the boldeft are " the beft. . . . Some Falcons are lazy and cow- " ardly; others are of fo fiery a temper, that " they can bear no reftraint; both thefe kinds " are to be rejected," &c.

Forget,

Forget, director of falconry at Verſailles, has been pleaſed to favour me with the following note :

" There is no material difference between the " Falcons of different countries, except in the " ſize; thoſe which come from the north are " commonly larger than thoſe from the moun- " tains, from the Alps, and Pyrenees ; the " latter are taken in the neſt, the former are " caught in their paſſage in various regions ; " they migrate in October and November, and " return in February and March. . . . The age " of Falcons is marked very preciſely in the ſe- " cond year, that is, at the firſt moulting ; but " afterwards it is much more difficult to diſ- " tinguiſh it. It may however be diſcovered till " the third moulting, not only from the changes " of the colour of the plumage, but from the " complexion of the feet and cere."

FOREIGN BIRDS,

WHICH ARE RELATED TO THE JER-FALCON AND FALCONS.

I.

THE Iceland Falcon, which, as we have already said, is a variety of the common species, and differs only in being rather larger and stronger.

II.

The Black Falcon is a bird of passage at Malta, in France, and in Germany, which Frisch and Edwards have figured and described, and which appears to us of a different species from our Common Falcon. I may observe, that the account given by Edwards is accurate, but that Frisch had no foundation for asserting that this Falcon is undoubtedly the strongest of all the rapacious birds that are of an equal size, because its upper mandible terminates in a sort of sharp tooth; and that it has larger toes and nails than the other Falcons; for we found upon comparison, that in regard to the toes and nails, it differed nothing from other Falcons; and in most of these the upper mandible had a similar termination; so that the marks of distinction which Frisch assigns are false or nugatory.

The

The Spotted Falcon, of which Edwards gives a figure and defcription, and which he tells us came from the fame country with the Black Falcon, that is, from Hudfon's Bay, appears to be only the Sorrel Falcon, or the young of the fame fpecies ; it owes its colours therefore to the difference of age, and not to any abfolute diftinction of kind. We have been affured that moft of the Black Falcons arrive from the fouth ; and yet we have feen one which was caught on the coaft of North America, near the banks of Newfoundland. Edwards tells us, that it is found in the country about Hudfon's Bay. We may therefore conclude that the fpecies is widely fcattered, and that it vifits alike the warm, the temperate, and the cold climates.

We may obferve, that in the bird which we faw, the feet were of a diftinct blue, while in thofe figured by Frifch and Edwards, they are yellow ; yet there is no doubt but the birds are the fame. We have noticed fome Ofpreys which had blue feet, and others which had them yellow ; this character is therefore not fo conftant as generally fuppofed. Indeed, like that of the plumage, it varies with the age, or with other circumftances.

III.

The bird which may be called the *Red Falcon of the Eaft Indies :* Aldrovandus * defcribes

* *Falco rubens Indicus.* ALDROV.

it

it accurately, and nearly as follows:—In the female, which is a third larger than the male, the upper part of the head is broad, and almoft flat: the colour of the head, neck, all the back and the upper part of the wings, is afh, verging on brown; the bill is very thick, though the hook is pretty fmall; the bafe of the bill is yellow, and the reft, as far as the hook, is cinereous; the pupil of the eyes is very black, the iris brown, the whole of the breaft, the higher part of the upper furface of the wings, the belly, the rump, and the thighs, are orange inclined to red; above the breaft and below the chin there is a long cinereous fpot, and feveral fmall fpots of the fame colour on the breaft; the tail is radiated with femicircular bars, alternately brown and afh-coloured; the legs and feet are yellow, and the nails black.   In the male all the parts which are red have a richer colour; thofe which are cinereous have more brown; the bill is bluer and the feet more yellow.   Thefe Falcons, Aldrovandus fays, were fent from India to the Grand Duke Ferdinand, who directed them to be delineated.—We may here obferve, that Tardif*, Albert, and Crefcent, have mentioned the Red Falcon as a fpecies or variety known in Europe, and inhabiting flat and marfhy countries: but this is not diftinctly enough defcribed for us to

---

* The Red Falcon is often found in flat fituations, and in marfhes; it is bold, but difficult to controul. TARDIF.

decide,

decide, whether it is the Eaſt Indian kind, which
might viſit Europe like the Paſſenger Falcon.

IV.

The bird mentioned by Willoughby under the
name of *The cirrated Indian Falcon*, which is
larger than the Common Falcon, and nearly of the ſize of the Gos-hawk; which has a
creſt divided at the extremity into two parts, that
are pendent on the neck. It is black on all
the upper parts of the head and body; but on
the breaſt and belly, the uniformity of colour
is interrupted by lines, which are alternately
black and white; the feathers of the tail rayed
with lines alternately black and cinereous; but
the feet are feathered to the toes; the iris, the
cere, and the feet, are yellow; the bill is of
a blackiſh blue, and the nails are of a fine
black.

In general it appears from the relations of
travellers, that the genus of the Falcons is one
of the moſt univerſally diſperſed. We have
already obſerved that it is found through the
whole extent of Europe, in the iſlands of the
Mediterranean, and on the ſhores of Barbary.
Dr. Shaw, whoſe narrative I find to be almoſt
always faithful and accurate, tells us, that in
the kingdom of Tunis there are Falcons and
Sparrow-hawks in abundance, and that they
form one of the principal amuſements of the
Arabs, and of the people of eaſier circumſtances.
They

They are ftill more common in the Mogul Empire *, and in Perfia †, where it is faid falconry

* " In Mogul the Falcon is flown at does and antelopes."
Voyage de Jean Ovington.

† The Perfians are expert in training birds for the chace; and
they generally inftruct the Falcons to fly at all forts of birds; and
for this purpofe they take cranes and other birds, and putting out
their eyes, they fet them at liberty, and immediately let loofe the
Falcon, which eafily catches them. . . . There are alfo Falcons for
the chace of antelopes : they are trained in the following manner :
They make the Falcons conftantly eat off the nofe of ftuffed antelopes, and fuffer them to feed no where elfe. After the birds are
thus bred, they carry them into the fields, and when they difcover an
antelope, they let loofe two of them, one of which faftens on the face
of the beaft, and ftrikes it before with the feet. The antelope ftops
fhort, and endeavours to fhake off the Falcon, which claps its wings
to keep its hold, and thus retards the flight of the antelope. When
after much ftruggling the Falcon is difengaged, another fucceeds;
and thus the antelope is continually haraffed and detained, until
the dogs have time to overtake it. Thefe fports are the more pleafant as the country is flat and open, and little interrupted by wood.
Relation de Thevenot. . . . *Voyage de Jean Ovington.*

The way in which the Perfians breed the Falcons to the chafe
of wild deer is, to fkin one and ftuff it with ftraw, and to faften the
flefh with which they feed Falcons always on the head of the ftuffed
animal, which is moved along on a four-wheeled vehicle in order to
accuftom the bird. . . . If the beaft is large, they fly feveral birds at
it, which teafe it one after another. . . . They alfo ufe thefe birds
in rivers and marfhes, into which they enter like dogs to hunt for
the game. . . . As all the military people are fportfmen, they ufually
carry at the pommel of the faddle a fmall tymbal of eight or nine
inches in diameter, and by ftriking it they recal the bird. *Voyage
de Chardin.*

Perfia has alfo birds of prey; there are many Falcons, Sparrow-
hawks, and Lanners, with which the royal venery is provided,
amounting to more than eight hundred. Some are flown at the
wild boar, the wild afs, and the antelope ; others are intended
againft cranes, herons, geefe, and partridges. A great part of
thefe birds of fport are brought from Ruffia ; but the largeft and
beft come from the mountains which ftretch towards the fouth from
Schyras unto the Gulph of Perfia. *Dampier's Voyage.*

is

is ſtudied with greater attention than in any other part of the globe *. They occur alſo in Japan, where Kœmpfer ſays they are brought from the northern parts of the iſlands, and are kept rather for oſtentation than utility. Kolben alſo makes mention of the Falcons at the Cape of Good Hope, and Boſman of thoſe on the coaſt of Guinea †. In ſhort, there is no part of the antient continent that is not ſtocked with Falcons ; and as they can ſupport cold, and fly with eaſe and rapidity, we need not be ſurpriſed to find them in the new world. Accordingly they have been diſcovered in Greenland ‡, in the mountainous traƈts of North and South America ‖, and even in the iſlands of the South Sea §.

* The Perſians, who have much perſeverance, take pleaſure in training a crow in the ſame way as the Sparrow-hawk. *Dampier's Voyage.*

† On the coaſt of Guinea there is another bird which reſembles much the Falcon, and which, though rather larger than a pigeon, is ſo bold and vigorous that it attacks and carries off the largeſt poultry. *William Boſman's Letters.*

‡ There are White and Grey Falcons in Greenland, where they are more numerous than in any other part of the world. Some of theſe birds were ſent as great rarities on account of their excellence to the kings of Denmark, who made preſents of them to other kings and princes their friends or allies, becauſe falconry is not praƈtiſed in Denmark, or in other parts of the north. *Recueil des Voyages du Nord.*

‖ Many Falcons of different ſpecies have been ſent from Mexico and Peru to the grandees of Spain, as they are highly valued. There are alſo herons and eagles of various kinds ; and no doubt theſe birds and others ſimilar could more eaſily migrate into thoſe countries than the lions and tigers. *D'Acosta's Nat. Hiſt. of the Weſt Indies.*

N. B. The bird which the Mexicans call *Hotli*, mentioned by Fernandez, appears to be the ſame with the Black Falcon, of which we have ſpoken.

§ *Hiſt. des Navig. aux Terres Auſtrales.*

The

V.

The bird called *Tamas* by the Negroes of
Senegal, and which was prefented to us by
Adanfon under the name of *Fifher-Falcon*. It
refembles the Common Falcon almoft entirely
in the colours of its plumage ; it is, however,
rather fmaller, and has on its head long erect
feathers, which are reflected back, and form a
fort of creft that diftinguifhes it from all others
of the fame genus. Its bill is yellow, not fo
much curved, and thicker than that of the Common Falcon, and its mandibles have confiderable indentations. Its inftinct is alfo different ;
for it fifhes rather than hunts. I imagine that
this is the fpecies which Dampier mentions by
the name of *Fifher-Falcon*. " It refembles," he
fays, " in colour and figure our fmaller forts of
" Falcons ; and its bill and talons are fhaped
" the fame. It perches upon the dry branches
" and trunks of trees that grow by the fides of
" creeks, rivers, or near the fea-fhore. When
" they obferve little fifh near them, they fkim
" along the furface of the water, feize them
" with their talons, and hurry them into the
" air without wetting their wings." He adds
" That they do not fwallow the fifh entire, like
" other birds that fubfift on that prey, but tear
" it with their bill, and eat it by morfels." [A]

[A] This is the *Falco Pifcator* of Gmelin and Latham. The
fpecific character :—" It is half-crefted, the head ferruginous, the
body cinereous, the quills have a dufky margin, the under fide of
the body pale yellowifh, with dufky longitudinal fpots."

THE HOBBY.

The HOBBY.

Le Hobreau, Buff.
Falco Subbuteo, Linn. Ray, Will. Aldr. &c.
Falco Barletta, Ger. Orn.
Dendrofalco, Briff. Frifch.
Baum-Falck, Gunth. Neft.

THE Hobby is much fmaller than the Falcon, and of a different difpofition. The fiery courage of the Falcon prompts him to attack birds that are far fuperior in fize; but the cautious Hobby, unlefs it is trained to the chace, never afpires beyond the prey of larks and quails. The want of boldnefs, however, is compenfated by its induftry. No fooner does it efpy the fportfman and his dog than it hovers in the train, and endeavours to catch the fmall birds that are put up before them; and what efcapes the fowling-piece eludes not the Hobby. It feems not intimidated by the noife of fire-arms, and ignorant of their fatal effects; for it continues to keep clofe to the perfon who fhoots. It frequents the champaign country near woods, efpecially where the larks are numerous. It commits great havoc among them, and thefe are well apprized of their fatal enemy; they are alarmed when they defcry it, and inftantly dive into the bufhes, or feek concealment in the herbage.

herbage. This is the only way in which the lark can effect its escape ; for though it soars to a great height, the Hobby can still outstrip it. The Hobby lodges and breeds in the forests, and perches upon the tallest trees. In some of our provinces the name of *Hobby* is applied to the petty barons who tyrannize over their peasants, and more particularly to gentlemen of the sport who chuse to hunt on their neighbours' grounds without obtaining leave, and who hunt less for pleasure than for profit *.

We may observe, that in this species the plumage is blacker during the first year than in the succeeding ones. In France there is a variety of the Hobby, which is represented Pl. Enl. N° 431. The difference consists in this ; that the throat, the lower part of the neck, the breast, a part of the belly, and the great feathers of the wings, are cinereous and without spots ; whereas, in the Common Hobby, the throat and the lower part of the neck are white, the breast and the upper part of the belly are white also, with longitudinal brown spots, and the great feathers of the wings are almost blackish. The tail, which in the common species is whitish below, dashed with brown, is in the variety entirely brown. But notwithstanding such

* This application of the name *Hobby* to country gentlemen might also be owing to another circumstance. Those who were not rich enough to keep Falcoss were contented with breeding Hobbies.

differ-

differences, thefe two birds are ftill of the fame
kind ; for their fize and port are the fame, and
they are both natives of France ; and befides,
they have in common a fingular character, that
the lower part of the belly and the thighs are
covered with feathers of a bright ruft-colour,
and which is ftrongly contrafted with the reft
of the plumage. It is even not unlikely that all
this diverfity of colours arifes from the age or
the feafon of moulting.—We have only to add,
that the Hobby is carried on the hand without
any cover or hood like the Merlin, the Sparrow-hawk, and Gos-hawk, and that it was formerly much ufed in hunting partridges and
quails. [A]

[A] The fpecific character of the Hobby, *Falco Subbuteo* : —
" The cere and feet yellow, the back dufky, the nape of the
" neck white, the abdomen pale with dufky oblong fpots, the
" under fide of the rump and the thighs rufous."

The male weighs feven ounces ; the length twelve inches ; the
alar extent two feet feven inches. It inhabits Europe and Siberia.
In fummer it is frequent in England, where it breeds, and migrates
in October.

It has been mentioned in the text, that the larks dread the fight
of the Hobby. They remain fixed to the ground through fear,
which affords the fowler an opportunity of fpreading his net over
them. This was formerly practifed, and termed *daring* the larks.

The German name *Baum-Falck* fignifies Tree-Falcon, which
Frifch has tranflated into a compound Greek and Latin term,
Dendrofalco.

The KESTREL*.

La Cresserelle, Buff.
Falco Tinnunculus, Linn. Ray, Will. Frif. &c.
Cenchris, Klein.
Falco Aureus, Id.
Bothel Geyer, Gunth.
Kirch Falck, Brunn.
Windwachel, Bittelweyer, Wannenweyer, Kram.
Gheppio Acertello, Gavinello, Zinn.
Stännel, Stonegall, Windhover, Alb. Sloan, &c.

THE Keſtrel is one of the moſt common of
the birds of prey in France, and eſpecially
in Burgundy.   There is ſcarcely an old caſtle or
deſerted tower, but is inhabited by it; and in the
mornings and evenings particularly it is ſeen
flying about the ruins.   It is ſtill oftener heard;
it conſtantly repeats, when on the wing, its quick
plĭ, plĭ, plĭ, or *prĭ, prĭ, prĭ*, and terrifies all the
ſmall birds, on which it ſhoots like an arrow,

---

* The Greek name Κεγχϱις, which ſignifies *millet*, is applied
to the Keſtrel, becauſe, as Geſner conjectures, the plumage of
this bird is ſprinkled with black ſpots like millet.   The Latin appellation *Tinnunculus* from *Tinnitus*, probably alludes to its tinkling
notes.   The former, *Rothel Geyer*, means Reddiſh Vulture; *Kirch
Falck*, Church Falcon; *Windwachel*, Wind-bird; *Rittelweyer*, Rider-
kite; *Wannen-weyer*, Fanner-kite;—the three laſt refer to the fan-
ning motion made by this bird.

In Italian it is alſo called *Tittinculo, Tintarello, Garinello, Canibello*.
In Spaniſh, *Cernicalo*, or *Zernicalo*.   It has been named in Engliſh,
the *Stonegall*, or *Stannel*, and the *Windhover*.

and

THE KESTRIL FALCON.

and feizes them with its talons; or if it miffes
the firft dart, it purfues them without fear even
to the houfes: I have known my fervants more
than once catch the Keftrel and its little fugitive,
by opening the window or the hall door, which
was more than one hundred fathoms from the
old walls where the purfuit commenced. After
it has fecured its prey, it kills it, and plucks the
feathers neatly; but it is not at fuch pains with
mice, for it fwallows the fmall ones entire, and
tears the large ones into pieces. The foft parts
of the carcafe are digefted in the ftomach of the
bird, but the fkin is rolled into a ball, and rejected at the bill. Its excrements are almoft
liquid, and whitifh; and the rolls that are thrown
out are found, by foaking in warm water, to be
the entire fkins of the mice. The Owls, Buzzards, and perhaps many other kinds of rapacious birds, reject alfo fimilar balls, which, befides the fkin, contain often the hardeft portions
of the bones. The fame is the cafe with Fifher-
birds; the bones and fcales of the fifhes are collected in the ftomach, and thrown out at the
bill.

The Keftrel is a pretty bird; its fight is acute,
its flight eafy and well fupported: it has perfe-
verance and courage, and refembles in its inftinct the noble and generous birds; and perhaps it might be trained, like the Merlins, for
falconry. The female is larger than the male:
its head is ruft-coloured, the upper fide of its

back, wings, and tail marked with crofs bars of brown, and all the feathers of the tail are of a rufty brown varioufly intenfe; but in the male, the head and tail are grey, and the upper parts of the back and wings are of a vinous ruft colour, fprinkled with a few fmall black fpots.

We cannot omit to obferve, that fome of our modern nomenclators have termed the female Keftrel the *Lark hawk (epervier des alouettes)*, and have reckoned it a diftinct fpecies from the Keftrel.

Though this bird habitually frequents old buildings, it breeds feldomer in thefe than in the woods: and when it depofits its eggs neither in the holes of walls nor in the cavities of trees, it conftructs a very flimfy fort of neft, compofed of fticks and roots, pretty much like that of the jays, upon the talleft trees of the foreft; fometimes it occupies the nefts deferted by the crows. It lays four eggs, but oftener five, and fometimes fix or feven; of which the two ends have a reddifh or yellowifh tinge fimilar to the plumage. Its young are at firft covered with a white down, and fed with infects; they are afterwards fupplied with plenty of field mice, which it defcries from aloft, as it hovers or wheels flowly round, and on which it inftantly darts. Sometimes it carries off a red partridge, which is much heavier than itfelf, and often catches pigeons that ftray from the flock. But, befides field mice and reptiles, its ordinary prey are fparrows, chaffinches, and

other

other ſmall birds. As it is more prolific than moſt
of the rapacious tribe, the ſpecies is more numerous and wider diffuſed ; it is found through
the whole extent of Europe, from Sweden to
Italy and Spain, and it occurs even in the more
temperate parts of North America. Many
Keſtrels continue the whole year in France ;
but I have obſerved that they are much leſs frequent in winter than in ſummer, which induces
me to think, that ſeveral migrate into other
countries to paſs the inclement ſeaſon.

I have raiſed numbers of theſe birds in large
volaries : they are, as I have already obſerved,
of a very fine white during the firſt month ;
after which the feathers on the back become
ruſty or brown in a few days : they are hardy,
and eaſy to feed ; they eat raw fleſh when it is
offered to them, when they are a fortnight or
three weeks old. They ſoon become acquainted
with the perſon who takes care of them, and
grow ſo tame as never to give offence : they
early acquire their cry, and repeat the ſame in
confinement as in the ſtate of liberty.—I have
often known them eſcape, and return of their
own accord after a day or two's abſence, probably compelled by hunger.

I am acquainted with no varieties of this ſpecies, except a few, in which the head and the
two feathers of the middle of the tail are gray,
ſuch as figured by Friſch ; but Salerne mentions
a yellow Keſtrel, which is found in Sologne,

Q 3 and

and of which the eggs are of the fame yellow hue. " This Keftrel," fays he, " is rare, and " fights nobly with the White John, which, " though ftronger, is often forced to yield the " conteft; they have been feen," he adds, " to hook together in the air, and fall to the " ground like a clod or a ftone." This appears to me very improbable; for not only is the White John much fuperior to the Keftrel in ftrength, but its movements are performed fo differently, that the birds could fcarcely ever meet. [A]

[A] The fpecific character of the Keftrel, *Falco Tinnunculus*, LINN. is, " that the cere and feet are yellowifh; the back rufous, " with black points; the breaft marked with dufky ftreaks; the " tail rounded."

It was formerly trained in Great Britain, to catch fmall birds and young partridges, but laid afide when falconry fell into difufe.

It is frequent in the deferts of Tartary and Siberia, and breeds in the fmall trees fcattered through the open country. It appears in Sweden early in the fpring, and departs in September. It is uncertain whether it penetrates farther north.

The STONE-FALCON.

Le Rochier, Buff.
Falco Lithofalco, Gmel. Briff. Will. &c.

THIS bird is not fo large as the Keftrel, and appears to me very like the Merlin, which is employed in falconry. It lodges and breeds, we are told, in rocks. Frifch is the only naturalift preceding us, who has given a diftinct defcription of it; and, upon a comparifon of his figure with thofe which we have given of the Keftrel and Merlin, we are much inclined to believe, that the Stone-falcon and the fpecies of the Merlin ufed in falconry are the fame, or at leaft clofely related :—but we fhall confider this more particularly in the following article. [A]

[A] The Linnæan character of the Stone-falcon: " The cere is " yellowifh, the upper fide of the body cinereous, the under-fide " rufous, with dufky longitudinal fpots, the tail cinereous, black " ifh near the end, the tip white."—It is about the fize of a Keftrel, and twelve inches and one-fourth long.

The MERLIN*.

L'Emerillon, Buff.
Falco Æsalon, Gmel. Ray. Will. Klein. Briss.
Cenchris, Fris.
Accipiter Smerillus, Ger. Orn.

THE subject of this article is not the Merlin of the naturalists, but that of the falconers, which has not been well described by any of our nomenclators. If we except the Butcher-bird, it is the smallest of all the rapacious tribe, not exceeding the size of a large thrush. Still we must reckon it a generous kind, and the nearest approaching the species of the Falcon: it has the same plumage †, the same shape and attitude, the same disposition and docility, and not inferior in ardour and courage. It can be successfully flown against larks, quails, and even partridges, which it seizes and carries off, though they are much heavier than itself; often

---

* The Latin name *æsalon*, is the same with the Greek αισαλων, which is of uncertain derivation. The modern names seem to allude to its preying upon blackbirds, *merulæ: Smerlo* or *Smerglio* in Italian; *Myrle* or *Smyrlin* in German; and *Merlin* in English. In some provinces of France, it is called the *Passetier*, or *Sparrow-catcher*.

† It actually resembles the Sorrel-Falcon in the shades and distribution of its colours.

it

THE MERLIN.

it kills them with one blow, ftriking on the
ftomach, head, or the neck.

This fmall bird, which refembles the Com-
mon-falcon fo much in its difpofition and courage *, is however fhaped more like the Hobby,
and ftill more like the Stone-falcon : but its wings
are much fhorter than thofe of the Hobby, and
reach not near the end of the tail; while, in the
Hobby, they project fomewhat beyond it.   We
have hinted in the preceding article, that its relation to the Stone-falcon is fo clear, in the thicknefs and length of the body, in the fhape of the bill,
feet, and talons, in the colours of the plumage, the
diftribution of the fpots, &c. that there is reafon
to fuppofe that the Stone-falcon is a variety of
the Merlin, or at leaft that they are two fpecies
fo nearly connected, that they ought to fufpend
any decifion refpecting their diverfity.—The
Merlin differs from the Falcons, and indeed all
the rapacious tribe, by a character which ap-
proximates it to the common clafs of birds ; viz.
the male and female are of the fame fize.   The
great inequality of fize therefore obferved between the fexes in birds of prey, cannot be attributed to the mode of life, or to any peculiar
habit : it would feem at firft to depend upon the
magnitude ; for, in the Butcher-birds, which are
ftill fmaller than the Merlins, the males and

* Many others have mentioned the analogy between the Merlin
and the Falcon; they have termed it the *Little Falcon*, *Falco par-*
vus Merlinus, SCHWENCK. and FALCONELLUS. RZACHYNSKI.

females

females are of the fame fize; while, in the
Eagles, the Vultures, the Jer-falcons, the Gos-
hawks, the Falcons, and the Sparrow-hawks,
the female is a third larger than the male.
Upon confulting the accounts of the diffection
of birds, I find that moft females have a large
double *cæcum*, while the males have only one
cæcum, and fometimes none at all : this difference
of the internal ftructure, which is much more
frequent in the females than in the males, is perhaps the true phyfical caufe of this exuberant
growth.   I fhall leave it to anatomifts to afcertain
the fact more accurately.

The Merlin flies low, though with great celerity and eafe : it frequents woods and bufhes
to feize the fmall birds, and hunts alone unaffifted by its female : it breeds in the mountain
forefts, and lays five or fix eggs.

But, befides the one we have juft defcribed,
there is another kind of Merlin better known
by naturalifts, which Frifch has figured and
Briffon defcribed from nature.   This differs confiderably from the former, and feems to refemble more the Keftrel ; at leaft, if we may judge
from the figure, not being able to procure a fpecimen.   But another circumftance feems to countenance this opinion : the American birds, which
we received by the name of the *Cayenne Merlin*
(Pl. Enl. N° 444), and the *St. Domingo Merlin*
(Pl. Enl. N° 465), appear to be varieties, or perhaps the male and female, of the fame fpecies,

4   and,

and, when viewed attentively, difcover more refemblance to the Keftrel than to the Merlin of
the falconers. This would imply, that the Keftrel
has migrated into the new continent; and accordingly, as a further prefumption, Linnæus
ranks it among the natives of Sweden, while he
omits the Merlin. We may therefore diftinguifh
it by a particular name, and that given it in the
Antilles may not be improper. "The Merlin,"
fays Father Tertre, " which our fettlers call *gry-*
" *gry*, from the cries which it makes in flying,
" is another fmall bird of prey that is fcarcely
" larger than a thrufh: all the feathers on the
" upper fide of the back and wings are rufty,
" fpotted with black; the under fide of the
" belly is white, fpeckled with ermine: it is
" armed with a bill and talons proportioned to
" its fize: it preys only on fmall lizards and
" grafs-hoppers, and fometimes on young chick-
" ens newly hatched: I have frequently refcued
" them; the hen makes a ftout defence, and
" drives off its enemy.—The fettlers eat it, but
" it is not very fat."

The refemblance between the cry * of this
Merlin of Father Tertre and that of the Keftrel,
is another mark of the proximity of thefe fpecies;
and it appears that we may conclude with tolerable certainty, that all the birds mentioned by

* The cry of the Keftrel is *prĭ, prĭ*, which is much like *gry-gry,*
the name of the bird of the Antilles.

naturalifts

naturalifts under the names of *Merlin of Europe*, *Carolina* or *Cayenne Merlin*, and the *St. Domingo Merlin*, or that *of the Antilles*, form only one variety in the fpecies of the Keftrel, and which we may diftinguifh from the common Keftrel by the appellation of *gry-gry*. [A]

[A] The fpecific charaéter of the Common Merlin, *Falco Æfalon* of Linnæus : "The cere and feet are yellow, the head "ferruginous, the upper fide of the body afh-cœrulean, with "ferruginous fpots and ftreaks, the underfide yellowifh white, "with oblong fpots." Gmelin and Latham regard the falconers' Merlin, and that of the Antilles, as fimple varieties.

The SHRIKES.

*Les Pie-Grieches**, Buff.

(Including the genus Lanius *in the Linnæan*
system.)

THOUGH these birds are small and of a delicate
make, yet their courage, their appetite for
carnage, and their large hooked bill, entitle them
to be ranked with the boldest and the most san-
guinary of the rapacious tribe: it is astonishing
with what intrepidity the little Shrikes combat
the Magpies, the Crows, and the Kestrels, which
are all much larger and stronger than themselves.
Not only do they act on the defensive, but they
sometimes commence the attack; and they are
ever successful in the rencounter, especially when
the parents unite to drive the birds of prey to a
distance from their nest. If they fly near their
retreats, the Shrikes rush upon them with loud
cries, inflict terrible wounds, and force them to
retire with little inclination to repeat the visit.
The more generous of the rapacious tribe regard
them with respect, and the Kites, the Buzzards,
and the Crows seem rather intimidated at their
appearance. Nothing in nature can give a better idea of the privileges annexed to courage,

* i. e. The Speckled Magpie.

than

than to fee thefe little birds, fcarcely equal in fize
to the larks, flying with fecurity among the
Sparrow-hawks, the Falcons, and other tyrants
of the air, and hunting in their domains without apprehending danger: for, though they
commonly live upon infects, they prefer flefh ;
they chafe all the fmall birds upon wing, and
they fometimes catch partridges and young hares.
Thrufhes, black-birds, and other birds caught
in the noofe, are their common prey ; they fix on
them with their talons, fplit the fkull with their
bill, fqueeze or cut the neck, and then pluck off
the feathers, and feed at their leifure, and tranfport the mangled fragments to their nefts.

The genus of thefe birds confifts of a vaft
number of fpecies ; but we may reduce thofe of
our climate to three principal ones : thefe are,
the Great Cinereous Shrike, the Woodchat, and
the Red-backed Shrike. Each of thefe three
fpecies requires a feparate defcription, and includes fome varieties which we fhall notice.

THE GREAT CINEREOUS SHRIKE

The Great CINEREOUS SHRIKE.

La Pie-Grieche Grise *, Buff.
Lanius Excubitor, Linn. Brun. Kram.
Falco Congener, Klein.
Lanius, feu *Collurio Cinereus Major,* Ray & Will. Briff. & Frif.
Ferlotta Berettina, Zinn.
Caſtrica Palombina, Olin.
Il Falconetti, Cett.
The Greater Butcher bird, or *Mattageſs* ; in the north of
England *Wierangle,* Will.
The Night Jar, Mort. North.
The Butcher bird, Murdering bird, or *Shreek* †, Mer. Pinax.

THIS bird is very common in France, where
it continues during the whole year. It inhabits the woods and mountains in ſummer, and
reſorts to the plains and near our dwellings in
winter. It breeds among the hills, either on the
ground or on the loftieſt trees. Its neſt is compoſed of white moſs interwoven with long graſs,
and well lined with wool, and is commonly
faſtened to the triple cleft of a branch. The
female, which differs not from the male in point
of ſize, and is only diſtinguiſhed by the lighter caſt
of its plumage, lays generally five or ſix eggs,

* *i. e.* The Grey Speckled Magpie.

† In modern Greek it is called *Collurio* ; the Latin name *Lanius*
ſignifies a *butcher.* In Italian it is termed *Gaza Sperviera, Fal-*
conello, Oreſto, Caſtrica, Verla, Stragazzina, Ragazzoia : in German, *Thorn-Kretzer, Warkengel, Nun-Maerder :* in Poliſh, *Zierzba,*
Strokos, Wiekſzy.

some-

fometimes feven or even eight, as large as thofe of
a thrufh. She feeds her young at firft with caterpillars and other infects, but foon inftructs them
to eat bits of flefh, which her mate brings with
wonderful care and attention. Very different
from the other birds of prey, which expel their
helplefs brood, the Shrike treats its infant young
with the moft tender affection, and even after
they are grown ftill retains its attachment. Towards autumn the offspring affift the parents in
providing for the common fupport; and the
members of the family continue during winter
to live in harmony, till the genial influence of
fpring awakens the appetite for propagation,
and forms other unions.

The Shrikes may be diftinguifhed both by
their flying in fmall troops after the breeding
feafon, and by their zig-zag courfe, which
waves not fideways, but bends with fudden
flexures upwards and downwards. They are
alfo difcovered by their fhrill cry *trŏuĭ*, *trŏuĭ*,
which can be heard at a great diftance, and
which they inceffantly repeat when perched on
the fummits of trees.

In this firft fpecies there is a variety in the
fize, and another in the colour. We have received for the King's cabinet a Shrike from Italy,
which differs from the common kind only by a
rufty tinge on the breaft and belly (Pl. Enl. Nº 32,
Fig. 1.). Some are found entirely white on the
Alps,

Alps *, which, as well as thofe with a rufous
tinge on the belly, are of the fame fize with the
Great Cinereous Shrike, and it does not exceed
the red-wing †.   But others are found in Germany and Switzerland which are fomewhat
larger, and which feveral naturalifts have reckoned a different fpecies ; yet in other refpects
thefe birds are fimilar, and their growth might
be affected by the plenty or fcarcity of fubfiftence which the country affords.   And if the
Great Cinereous Shrike varies fomewhat in
Europe, we may expect it to vary ftill more
in remote climates.   That of Louifiana (Fig. 2,
N° 476, Pl. Enl.) is the fame with the common
kind, differing lefs than the Italian bird ; only it
is rather fmaller, and of a deeper caft on the
upper parts of the body.   Thofe from the Cape
of Good Hope ‡ (Fig. 1, N° 477), and Senegal
(Fig.

---

* *Lanius Albus,* ALDROV.   This is the fecond variety of Linnæus :—" Its body white ; its feet yellowifh ; the bill and nails
blackifh."

† *Lanius Major,* GESNER ; and the *Groeferer Neuntoeder* of
FRISCH.   It is the third variety of *Lanius Excubitor,* LINN.   It
is larger and thicker than the former ; the fcapular feathers, and
the fmall coverts of the upper fide of the wing, are rufty-coloured :
But thefe differences are too minute to conftitute a feparate fpecies.

‡ To this fpecies we muft alfo refer the Eaft-India bird which
the Englifh that vifit the coafts of Bengal term the *Dial-bird,*
and which is defcribed by Albin with figures of the cock and hen.
" This Shrike is very large, he fays, and very fimilar to the Great
Cinereous Shrike ; its bill black, the corners of the mouth yellow,
the iris of the fame colour, the legs and feet brown.   In the male,
the head, the neck, the back, the rump, and the coverts of the
upper fide of the tail, the fcapular feathers, the throat, and the

(Fig. 1. N° 297), and the Blue Shrike from Madagascar (Fig. 1, N° 298), appear to be three contiguous varieties, and equally related to the Great Cinereous Shrike of Europe. The only differences are, that in the one from the Cape, the upper parts of the body are of a blackish brown; in that from Senegal, they are of a lighter brown; and in that from Madagascar, they are of a fine blue: but such differences of plumage may still have place in the same species, for we shall have frequent occasion to point out as great changes produced in our own climates, and the variations ought to be still greater in distant regions. The Shrike from Louisiana resembles that of Italy; and the temperature of these countries are nearly alike. The others, from the Cape, Senegal, and Madagascar, bear less analogy; and the climates to which they belong are also more different.—The Shrike from Cayenne is variegated with long brown bars (Pl. Enl. N° 297); but the size and other properties being the same, we have ranged it likewise with the common kind. [A]

breast, are black; the belly, the flanks, and the coverts of the under-side of the tail, white; all the feathers of the tail of an equal length, black above and white beneath. The female is distinguished from the male by its fainter colours.''

[A] Specific character of the Great Cinereous Shrike, *Lanius Excubitor*, Linn. " The tail wedge-shaped, its lateral quills white; the back hoary; the wings black, with a white spot." Its length is ten inches, its breadth across the wings fourteen inches, and it weighs three ounces. It seizes small birds by the throat, and
strangles

ftrangles them ; then fpits them on fome thorn, and tears them to pieces with its bill. Even when confined in a cage, it fticks its meat againft the wires, and tears it in the fame manner.—It is frequent in Ruffia, but feems not to have penetrated to Siberia. It inhabits the whole extent of North America. In Hudfon's-bay it breeds in the woods diftant from the coaft. It makes its neft with dry grafs, which it lines thick with feathers.

The WOODCHAT.

La Pie-Grieche Rouſſe *, Buff.
Lanius Rutilus, Lath.
Lanius Rufus, Briſſ. and Gmel.
Lanius Pomeranus, Muſ. Carlſc.
Lanius Minor Cineraceus, Ray, Klein, Friſ.
Ampelis Dorſo Griſeo, Faun. Suec. ed. 1. and Kram.
Buſerola, Ferlotta Bianca, Zinn.

THIS bird is rather ſmaller than the preceding, and may eaſily be diſtinguiſhed by
the tinge of its head, which is ſometimes red,
and commonly bright ferruginous ; its eyes alſo
are whitiſh or yellowiſh, while in the Great Cinereous Shrike they are brown, and its bill and
legs are blacker. Its inſtincts, however, are nearly the ſame ; both of them are bold and miſ-
chievous, yet they are evidently of different ſpecies ; for the Great Cinereous Shrike is a permanent ſettler, while the Woodchat quits the
country in autumn, and returns not till ſpring.
The family, which does not diſperſe after the
young are fledged, departs alone in the beginning of September ; they flutter from tree to
tree, and ſupport not a continued flight even in
their migrations. They reſide during ſummer
in the plains, and neſtle on the buſhy trees ; in

* *i. e.* The Rufous Speckled Magpie.

that

that feafon the Great Cinereous Shrike inhabits the forefts, and feldom emerges from the retreat till after the departure of the Woodchat. The Woodchat is faid to be the moft palatable of all the Shrikes, or perhaps the only one that is fit to be eaten *.

The male and female are almoft exactly of the fame fize, but differ fo much in their colours as to appear of diftinct fpecies. I fhall only add, that both the Woodchat and the Red-backed Shrike conftruct their neft very neatly, and employ the fame materials as the Great Cinereous Shrike; the mofs and wool are fo well connected with fmall pliant roots, long fine grafs, and the tender fhoots of low fhrubs, that the whole feems interwoven. It has generally five or fix eggs, fometimes more; thefe are of a whitifh ground, and either entirely or partly fpotted with brown or fulvous. [A]

* Schwenckfeld.

[A] The Woodchat, *Lanius Rutilus* of LATHAM, is thus-defcribed:—" Its upper fide confifts of three colours, its under fide is rufous-white; the whole of the fcapular feathers, the quills of the tail from the bafe to the middle, and the iris of the lateral ones, white; and a black ftreak through the eyes." It includes the *Lanius Rufus* of Gmelin, which is the third variety of the *Lanius Collurio* of Linnæus, or the Red-backed Shrike; and alfo the *Lanius Pomeranus*, firft defcribed by Sparmann."

The RED-BACKED SHRIKE.

L'Ecorcheur, Buff.
Lanius-Collurio, Linn. Gmel. Briff. Brun. Kram, &c.
Lanius Minor Rufus, Ray and Will.
Lanius Æruginofus Major, Klein.
Ferlotta Roffa, Zinn.
The Leffer Butcher bird, called in Yorkfhire *Flufher,* Will.

THE Red-backed Shrike is only a little
fmaller than the Woodchat; and its habits
are fimilar. It departs with its family in September, and returns in the fpring. It breeds in
the trees or bufhes in the open country, and not
in the woods. It feeds its young commonly
with infects, and preys upon the fmall birds.
In fhort, the only material difference confifts in
the fize, and in the diftribution and fhades of
the colours, which feem to be invariably difcriminated in both fpecies: but the difference is
ftill greater between the male and female in
each fpecies. We may therefore with propriety
regard the Woodchat, the Red-backed Shrike,
and the variegated Red-backed Shrike, as varieties of the fame fpecies. Some naturalifts* have
indeed reckoned the laft a diftinct fpecies; but
the comparifon of the figures feems to prove
that it is only the female of the Red-backed
Shrike.

* *Collurionis parvi fecundum genus.* ALDROV. *Collurio varius.*
BRISS.

Thefe

THE RED BACKED SHRIKE.

These two species of Shrikes, with their varieties, breed in Sweden as well as in France. We may presume therefore that they will be found in the new continent; and we may reckon the foreign kinds as only varieties of the Woodchat produced by the influence of climate.

Nothing can shew better that birds migrate into warmer countries to pass the winter, than the Woodchat (N° 477, Fig. 2, Pl. Enl.) sent by Adanson from Senegal, and which is precisely the same with the European Woodchat. There is another (N° 279, Pl. Enl.) which we received from the same place, and which may be regarded as merely a variety, since the only difference is, that its head is black, and its tail rather longer, which is not material.

The same observation may be extended to what we have called the *Philippine Woodchat** (Pl. Enl. N° 476, Fig. 1.), and also to the *Louisiana Shrike* (Pl. Enl. N° 397), which, though brought from climates widely different, appear to be really the

* It would appear that this bird is the same with what Edwards has called the *Red* or *Rufous Crested Shrike*. " This bird, says he, is termed Charah in the country of Bengal, and differs from our Shrikes by its crest.'' But this difference is slight; for what Edwards takes for a crest is only the feathers bristled, as in the Jay when irritated. He confesses, that he only saw the dried specimen; and what evinces our position is, that the same naturalist gives a figure of the Black and White Shrike of Surinam in the first part of his Gleanings, where it is represented with a crest; yet we have that species in the King's cabinet, and it undoubtedly is not furnished with a crest. Edwards was therefore misled by some accidental derangement of the feathers; and we may still assert that the Bengal Shrike is only a variety of the Red-backed Shrike.

same

fame bird, and only a variety of the Red-backed
Shrike, whofe female it refembles almoft exactly. [A]

[A] Specific character of the Red-backed Shrike, *Lanius-Collurio*, Linn. " Its tail fomewhat of a wedge-fhape, the back
grey, the four middle quills of the tail of an uniform colour, the
bill lead-coloured." It is feven inches and a half long, its alar
extent eleven inches ; the male weighs two ounces, the female
two ounces and two grains ; it inhabits Europe, and is migratory,
appearing in May, and returning in September or October.

FOREIGN BIRDS,

RELATED TO THE GREAT CINEREOUS AND RED-BACKED SHRIKES.

I.

The FORK-TAILED SHRIKE*.

Le Fingah, Buff.
Lanius Cærulescens, Linn.
Lanius Bengalensis Cauda Bifurca, Briff. and Klein.
The Forked-tail Butcher-bird, Edw.

EDWARDS speaks of this bird in the following terms:—The shape of its bill, the whiskers at its base, and the strength of its legs, have induced me to range it with the Shrikes; though its tail is different, being forked, while that of the Shrike has the longest feathers in the middle. Its bill is strong, thick, and arched, nearly like that of a Sparrow-hawk, but longer in proportion to its thickness, less hooked, and with wide nostrils. The base of the upper mandible is beset with stiff hairs. . . . The whole of the head, neck, back, and the coverts of the wings, are of a shining black, with reflections of blue, purple, and green, varying according to its position. . . . The breast is of an ash-colour, dusky,

* The specific character :—" The tail forked, the body cœrulean black, the abdomen white."

and

and blackiſh.   All the belly, the legs, and the
coverts of the under ſide of the tail, are white;
the legs, the feet, and the nails, are blackiſh
brown.—I am at a loſs, ſubjoins Edwards, whether I ſhould claſs this bird with the Shrikes or
the magpies; for it appears to be equally related to each of them; and I am even inclined
to think that both conſtitute only one ſpecies.—
This conformity ſeems to have been obſerved in
France, where the name *pie* is applied equally
to the Shrikes and magpies.

II.

The BENGAL SHRIKE*.

Rouge-Queue, Buff.
Lanius-Emeria, Linn. and Gmel.
Lanius Bengalenſis Fuſcus, Briſſ.
The Indian Redſtart, Edw.
The Bengal Redſtart, Alb.

This is alſo an Eaſt-India bird.   It is deſcribed and figured by Albin.   It is nearly of the
ſame ſize as the Great Cinereous Shrike of Europe; its bill is cinereous brown; its iris whitiſh; the upper part, and the back of the head,
black; below the eyes is a lively orange ſpot
terminated with white; and on the tail four
black ſpots, making a ſegment of a circle; the

* The ſpecific character:—" It is grey, white beneath, the
temples and rump red."   It is five inches and a half long.

upper

upper part of the neck, the back, the rump, the
fuperior coverts of the tail, the inferior coverts
of the wings, and the fcapular feathers, are
brown ; the throat, the upper part of the neck,
the breaft, the higher part of the belly, and the
inferior coverts of the tail, are red ; the tail is
light brown ; the feet and nails are black.

***

III.

LANGARIEN and TCHA-CHERT.

The bird fent from Manilla under the name
of *Langarien* *, and the other from Madagafcar
under that of *Tcha-chert* †, have perhaps been
improperly referred to the genus of Shrikes ; for
their wings extend beyond the tail ; while, in
the other fpecies, they do not reach fo far as the
tail. But the one from Madagafcar refembles
much our Great Cinereous Shrike ; and, fetting
afide the difference of the length of the wings,
we may confider it as the intermediate fhade
between that and the Manilla bird, to which
however it is nearer related ; and as we know

* The *Langarien* is the *Lanius Leucorynchos* of GMELIN, and
the *White-billed Shrike* of LATHAM. Its fpecific charaćter :—
" Blackifh above, whicifh below; the bill, breaft, abdomen, and
rump, white." It is feven inches long.

† The *Cha-Chert* is the *Lanius Viridis* of GMELIN, and the
Green Shrike of LATHAM. Specific charaćter :— " The upper
furface of the head, body, and wings, dull green ; the body white
beneath, the tail black." The length near fix inches.

2

no other genus to which we could directly refer them, we shall follow the opinion of the rest
of the naturalists, remarking at the same time
the uncertainty of the determination.

IV.

The CAYENNE SHRIKE.

Becarde *, Buff.
Lanius Cayanus †, Linn. Gmel. and Briss.

Two of these birds were sent ; the one under
the name of the Grey Shrike, the other under
that of the Spotted Shrike.   Their bill is large
and red ; their head is entirely black ; and their
size exceeds that of the European Shrikes, though
they resemble these on the whole more than any
birds of our latitudes.   They seem to be the male
and female of the same species.

   * So called on account of the thickness and length of the bill
(*bec*).
   † The specific character :—" Cinereous, the head, the quills of
the wings, and the primaries of the tail, black."   It is of the size
of the blackbird, being eight inches and a half long.

V.

The YELLOW-BELLIED SHRIKE*.

Becarde a Ventre Jaune, Buff.
Lanius Sulphuratus, Linn. and Gmel.
Lanius Cayanensis Luteus, Briss.

This bird has a long bill like the preceding, and therefore related to it. Indeed, the only difference consists in the colours of the plumage.

* Specific character : — " Dusky ; yellow beneath, the head blackish, and encircled by a whitish stripe." It is near nine inches in length.

VI.

The HOOK-BILLED SHRIKE†.

Le Vanga, ou *Becarde a Ventre Jaune*, Buff.
Lanius Curvirostris, Linn. and Gmel.
Collurio Madagascariensis, Briss.
Lanius Major Nigro & Albo Mixtus, Gerin. Orn.

This bird was sent from Madagascar by Poivre, under the name of *Vanga*. Though considerably different from the Shrikes, it seems to be more related to them than any other birds of Europe :—It bears a resemblance to the two preceding.

† Specific character : —" The tail wedge-shape, the body white, the back black, the five first quills of the wings marked with a white spot." It feeds upon fruits. It is ten inches long.

VII.
The RUFOUS SHRIKE*.

Schet-bé, Buff.
Lanius Rufus, Linn. and Gmel.
Lanius Madagaſcarenſis Rufus, Briſſ.

This was alſo ſent from Madagaſcar by Poivre:
—It is much like the preceding, and, did not the
diſtance of the countries preclude the idea, we
might ſuppoſe them to conſtitute the ſame ſpe‑
cies.   The Rufous Shrike is leſs removed from
the European Shrikes, than thoſe of Cayenne, for
its bill is ſhorter.

* Specific character:—" Rufous, white beneath, the head
greeniſh-black."   It is about eight inches long:

———————

VIII.
The WHITE-HEADED SHRIKE †.

Tcha-Chert-Bé, Buff.
Lanius Leucocephalus, Gmel.
Lanius Madagaſcarienſis Major Viridis, Briſſ.

We received this bird by the ſame channel:—
It ſeems to be a proximate ſpecies of the pre‑

† Specific character:—" Head white, the upperſide of the body
is a greeniſh-black, and beneath black; the bill, feet, and nails
lead-coloured."   It is eight inches long.

ceding,

ceding, or perhaps a variety of age or fex, its bill only being fomewhat fhorter and lefs hooked, and its colours rather differently diftributed. Indeed all thefe five birds with thick bills might form a fmall feparate genus.

IX.

The BARBARY SHRIKE*.

Le Gonolek, Buff.
Lanius Barbarus, Linn. and Gmel.
Lanius Senegalenfis Ruber, Briff.

We received this bird from Senegal, where the Negroes, as Adanfon informs us, call it *gonolek,* that is, *feeder on infects.* It is painted with the moft vivid colours: it is nearly of the fame fize as the European Shrike, and fcarcely differs in any thing but the diftribution of its tints, which is however nearly fimilar to what has place in the Great Cinereous Shrike of Europe.

* Specific character:—" Black, beneath red, the crown and thighs fulvous." It is about nine inches long.

X.

The MADAGASCAR SHRIKE*.

Lanius Madagascarensis, Linn. Gmel. and Briss.

Both the male and female of this bird were
sent from Madagascar by Poivre ; the former
under the name of *Cali-calic,* and the latter under that of *Bruia.* We might, on account of
its smallness, refer it to the genus of the European Red-backed Shrike ; but it differs so
much, that it ought to be regarded as a distinct species.

* Specific character:—" Cinereous, beneath whitish, the lines
between the bill and the eyes black, the quills of the wings
tawny." It is five inches long, and about the size of a
sparrow.

XI.

The CRESTED SHRIKE†.

Pie-Grièche Huppée.
Lanius Canadensis.

This bird, which was brought from Canada,
has on the crown of its head a soft crest, with long
feathers that fall backwards. It is similar to our
Woodchat in the distribution of its colours, and
may be regarded as a contiguous species, differing
scarcely in any thing but the crest and the bill,
which is rather thicker.

† Specific character :—" The tail is wedge-shaped, the head
crested, the body tawny, and below waved with fulvous, and
dusky." It is six inches and a half long. Sometimes wants the
crest.

The NOCTURNAL BIRDS of PREY.

THE eyes of thefe birds are fo delicate, that they feem to be dazzled by the fplendor of day, and entirely overpowered by the luftre of the folar rays; they require a gentler light, fuch as prevails at the dawn, or in the evening fhades. They leave their retreats to hunt, or rather to fearch for their prey, and their expeditions are performed with great advantage; for in this ftill feafon, the other birds and fmall animals feel the foft influence of fleep, or are about to yield to its foothing power. Thofe nights that are cheered by the prefence of the moon, are to them the fineft of days, days of pleafure and of abundance, in which they feek their prey for feveral hours together, and procure an ample fupply of provifions. When fhe withholds her filver beams, their nights are not fortunate; and their ravages are confined to a fingle hour in the morning and in the evening; for we cannot fuppofe that thefe birds, though they can diftinguifh objects nicely in a weak light, are able to perform their motions when involved in total darknefs. Their fight fails when the gloom of night is completely fettled; and in this refpect they differ not from other animals, fuch as hares, wolves, and ftags, which leave the woods in the evening to feed or to hunt during the night. only, thefe animals fee ftill better in the day than in the night; whereas the organs of vifion in the nocturnal birds are fo much overpowered by the brightnefs of the day, that they are obliged to remain in the fame fpot without ftirring; and when they are forced to leave their retreat, their flight is tardy and interrupted, being afraid of ftriking againft the intervening obftacles. The other birds, perceiving their fear, or their conftrained fituation, delight to infult them: the tit-moufe, the finch, the red-breaft, the black-bird, the jay, the thrufh, &c. affemble to enjoy the fport. The bird of night remains perched upon a branch, motionlefs and confounded, hears their movements and their cries, which are inceffantly repeated, becaufe it anfwers them only with infignificant geftures, turning round its head, its eyes, and its body with a foolifh air. It even fuffers itfelf to be affaulted without making refiftance; the fmalleft, the weakeft of its enemies are the moft eager to torment it the moft, determined to turn it into ridicule. Upon this play of mockery, or of natural antipathy, is founded the pretty art of bird-calling. We have only to put an Owl, or even to imitate its notes, in the place where the limed twigs are fpread, in order to draw the other birds.

13

birds*. The beſt time is about an hour before the cloſe of the day; for if this diverſion be deferred later, the ſame ſmall birds which aſſemble in the day to inſult over the bird of night with ſo much audacity and obſtinacy, avoid the rencounter after the evening ſhades have reſtored his vigour, and encouraged his exertions.

All this muſt be underſtood with certain reſtrictions, which it will be proper to ſtate here: 1. All the ſpecies of Owls are not alike dazzled with the light; the Great-eared Owl ſees ſo diſtinctly in open day, as to be able to fly to conſiderable diſtances; the Little Owl chaces and catches its prey long before the ſetting, and after the riſing of the ſun. Travellers inform us, that the Great-eared Owl or Eagle-Owl of North America catches the white grous in open day, and even when the reflection of the ſnow adds to the intenſity of the light: Belon remarks, that " whoever will examine the ſight of theſe birds, " will find it not ſo weak as is commonly ima- " gined." 2. It appears that the Long-eared Owl ſees worſe than the Scops, and is the moſt dazzled by the light of day, as are alſo the Tawny Owl, the White, and the Aluco; for theſe equally attract the ſame birds, and afford

* This ſort of ſport was known to the antients; for Ariſtotle diſtinctly mentions it in the following terms : " In the day all the other ſmall birds flock round the Owl, to admire it, as it is called, and ſtrike it. Whence, if it is ſet in a proper place, many ſort of ſmall birds may be caught."

them

them fport.   But before we relate the facts which
apply to each particular fpecies, we muft mention the general diftinctions.

The Nocturnal Birds of Prey may be divided
into two principal *genera*: the genus of the
Hibou, (the Long-eared or Horned Owl,) and
that of the *Chouette*, (the Earlefs or Little Owl,)
each of which contains feveral different fpecies.
The diftinguifhing character of thefe two *genera* is, that all the *Hibous* have two tufts of
feathers in the fhape of ears erect on each fide
of the head; while, in the *Chouettes*, the head
is round without tufts or prominent feathers *.
We fhall reduce the fpecies contained in the
genus of the *Hibou* to three.   Thefe are, 1.
The Great-eared Owl.   2. The Long-eared.
3. The Scops.   But the genus of *Chouette* includes at leaft five fpecies: which are, 1. The
Aluco.   2. The Tawny.   3. The White.   4.
The Brown. And 5. The Little Owl. Thefe eight
fpecies are all found in Europe, and even in
France; fome are fubject to varieties, which
feem to depend on the difference of climates;
others occur that refemble them in the New

* Pliny feems to have remarked this diftinction : " Of the feathered race, the *Bubo* and the *Otus* alone have feathers like ears."
Lib. xi. 37. And again, " The *Otis* is fmaller than the *Bubo*, larger
than the *Noctuæ*, and has feathers projecting from the ears, whence
its name; fome call it in Latin *Afio*." Lib. x. 23. N.B. There
are three fpecies with tufted ears : the Great-eared Owl *(Bubo)*; the
Long-eared Owl *(Otus)*; and the Scops-eared Owl *(Afio)*, which
Pliny confounds with the *Otus*.

World;

World; and indeed, moſt of the Owls of America differ ſo little from thoſe of Europe, that we may refer them to the ſame origin.

Ariſtotle mentions twelve ſpecies of birds which ſee in the dark, and fly during the night; and as in theſe he includes the Oſprey and Goat-ſucker, under the names of *Phinis* and *Ægotilas;* and three others, under the names of *Capriceps, Chalcis,* and *Charadrios,* which feed on fiſh, and inhabit marſhes, or the margins of lakes and ſtreams, it appears that he has reduced all the Owls known in Greece in his time to ſeven ſpecies: the Long eared, which he calls Ωτος, *otus,* precedes and conducts the quails when they begin their migration, and for this reaſon it is named *dux,* or *leader;* the etymology ſeems certain, but the fact muſt be ſuſpected. It is true that the quails, when they leave us in the autumn, are exceſſively fat, and ſcarcely fly but in the night, and repoſe during the day in the ſhade to avoid the heat; and hence the Long-eared Owl may ſometimes be obſerved to accompany or go before theſe flocks of quails; but it has never been obſerved that the Long-eared Owl is, like the quail, a bird of paſſage. The only fact which I have found in travellers that ſeems to countenance this opinion, is in the Preface to Cateſby's Natural Hiſtory of Carolina. He ſays, that in the twenty-ſixth degree of north latitude, being nearly in the middle of the Atlantic, in his paſſage to Carolina, he ſaw an Owl over the

veſſel;

veſſel ; and he was more ſurpriſed at this circumſtance, as that bird has ſhort wings, and is
eaſily fatigued. He adds, that after making
ſeveral attempts to alight, it diſappeared.

It may be alleged in ſupport of this fact, that
the Owls have not all ſhort wings, ſince in moſt
of theſe birds they ſtretch beyond the point of
the tail, and the Great-eared Owl and the Scops
are the only ſpecies whoſe wings do not reach
quite its length. Beſides, we learn from their
ſcreams, that all theſe birds perform long
journies ; whence it ſeems that the power of
flying to a diſtance during the night, belongs to
them as well as to the others ; but their ſight being leſs perfect, and not being able to deſcry remote objects, they cannot form an idea of a
great extent of country, and therefore have not,
like moſt other birds, the inſtinct of migration.
At leaſt, it appears that our Owls are ſtationary.
I have received all the ſpecies not only in ſummer, in ſpring, in autumn, but even in the
moſt piercing colds of winter. The Scops alone
diſappears in this ſeaſon ; and I have actually
been informed, that this ſmall ſpecies departs in
the autumn, and arrives in the ſpring : hence
we ought to aſcribe to the Scops, rather than the
Long eared Owl, the buſineſs of leading the
quails. But this fact is not proved, and I know
not the foundation of another fact advanced by
Ariſtotle, who ſays, that the Tawny Owl *(Glaux
Noctua,* according to his tranſlator Gaza) con-

ceals

ceals itfelf for feveral days together; for in the chilleft feafon of the year I have received fome that were caught in the woods: and if it be pretended that the words *Glaux Noctua* fignifies the White Owl, the fact would be ftill wider off the truth; for except in very dark and rainy evenings, it is conftantly heard through the whole year to whiftle and fcream about twilight.

The twelve Nocturnal Birds mentioned by Ariftotle, are: 1. *Byas;* 2. *Otos;* 3. *Scops;* 4. *Phinis;* 5. *Ægotilas;* 6. *Eleos;* 7. *Nyctico-rax;* 8. *Ægolios;* 9. *Glaux;* 10. *Charadrios;* 11. *Chalcis;* 12. *Ægocephalus;* which Theodore Gaza tranflates by the Latin words, 1. *Bubo;* 2. *Otus;* 3. *Afio;* 4. *Offifraga;* 5. *Caprimulgus;* 6. *Aluco;* 7. *Cicunia, Cicuma, Ulula;* 8. *Ulula;* 9. *Noctua;* 10. *Charadrius;* 11. *Chalcis;* 12. *Ca-priceps.*

The nine firft feem to be as follow :—

1. The Great eared Owl; 2. The Long-eared Owl; 3. The Scops; 4. The Ofprey; 5. The Goat-fucker; 6. The White Owl; 7. The Aluco Owl; 8. The Brown Owl; 9. The Tawny Owl.

All the naturalifts and men of letters will readi-ly admit that, 1. The *Byas* of the Greeks, *Bubo* of the Latins, is our Great-eared Owl. 2. That the *Otos* of the Greeks, *Otus* of the Latins, is our Long-eared Owl. 3. The name of Scops in the Greek, in Latin *Afio*, is the Small Owl. 4. The *Phinis* of the Greeks, *Offifraga* of the Latins,

is

is the Ofprey. 5. The *Ægotilas* of the Greeks,
Caprimulgus of the Latins, is the Goat-fucker.
6. That the *Eleos* of the Greeks, *Aluco* of the
Latins, is the White Owl. But at the fame time
it will be afked, why I tranflated *Glaux*, by the
Tawny Owl; *Nycticorax*, by *Aluco* ; and the
Ægolios, by the Brown Owl; while all the nomenclators and naturalifts who have preceded
me have rendered *Ægolios* by *Hulotte (Aluco)*,
and are obliged to confefs that they know not to
what bird to refer the *Nycticorax*, nor the *Charadrios*, the Chalcis, and the *Capriceps*. I fhall be
blamed for transferring the name of *Glaux* to the
Tawny Owl, fince it has been applied, by the uniform confent of all who have gone before me, to
the Brown Owl, or even to the Little Owl.

I proceed to explain the reafons which have
induced me to make thefe innovations, and to
remove the obfcurity which attends their doubts
and their falfe interpretations.   Among the
Nocturnal birds which we have enumerated, the
Tawny Owl is the only one whofe eyes are
blueifh, the Aluco the only one whofe eyes are
blackifh ; in all the reft the iris is of a golden,
or at leaft of a faffron colour.   But the Greeks,
whofe accuracy and precifion of ideas I have
often admired in the names which they have
applied to the objects in nature, which always
mark the characters in a ftriking manner, would
have had no foundation to beftow the name of
Glaux (glaucous, cærulean) upon birds which have
none

none of the blue fhade, and whofe eyes are black, orange, or yellow; but they would have had the beft reafon to give this name to that fingle fpecies which is diftinguifhed from all the reft by the blue tinge of its eyes; nor would they have called thofe birds whofe eyes are yellow or blue, and whofe plumage is white or grey, and bear no refemblance to the Raven, by the term *Nycticorax*, or Raven of night; but they would with great propriety have beftowed this name on that Owl, which is the only one of the Nocturnal Birds whofe eyes are black, and whofe plumage is almoft black, and which in its fize bears a greater analogy than any other to the Raven.

The probability of this interpretation derives additional force from another confideration. The *Nycticorax* was a common and noted bird among the Greeks, and even among the Hebrews, fince it is often the fubject of their comparifons *(ficut nycticorax in domicilio)*. We cannot imagine with thofe literati, that this bird was fo folitary and fo rare, that it can be no longer found. The Aluco is common in every country, it is the largeft of the Earlefs Owls; the blackeft and the likeft the Raven: it differs widely from every other fpecies; and this obfervation drawn from the fact itfelf, ought to have more weight than the authority of thofe nomenclators, who are too little acquainted with nature to interpret with accuracy its hiftory.

10

But

But admitting that the *Glaux* fignifies Tawny
Owl, the Earlefs Owl with blue eyes, and
Nycticorax, Aluco, or Earlefs Owl with white
eyes, the *Ægolios* muft be the Earlefs Owl with
yellow eyes.—This requires fome difcuffion.

Theodore Gaza renders the word *Nycticorax*,
firft by *Cicuma*, then by *Ulula*, and afterwards
by *Cicunia :* this laft is probably the miftake of
the tranfcribers, who have written *Cicunia* inftead of *Cicuma* ; for Feftus, prior to Gaza, alfo
tranflated *Nycticorax* by *Cicuma ;* and Ifidorus
by *Cecuma* ; others by *Cecua.*—To thefe words
we may even refer the etymology of *Zueta* in
Italian, and *Chouette* in French. If Gaza had
attended to the characters of the *Nycticorax*, he
would have adhered to his firft interpretation,
Ulula, and would not have made a double application of this term ; for he would, in that
cafe, have tranflated *Ægolios* by *Cicuma.* Upon
the whole therefore we may conclude, that
Glaux is the Tawny Owl, *Nycticorax* the Aluco
Owl, and *Ægolios* the Brown Owl.

The *Charadrios*, the *Chalcis*, and the *Capriceps*,
ftill remain to be confidered : Gaza is contented
with giving the Greek words a Latin termination.   But as thefe birds are different from thofe
of which we at prefent treat, and feem to be the
inhabitants of marfhes and the margins of lakes ;
we fhall defer the confideration of the fubject
till we come to the hiftory of the birds that fifh
in the twilight.   The Little Owl is the only

fpecies

species whose name I cannot discover in the Greek language. Aristotle never mentions it, and probably he confounded it with the Scops, which it indeed resembles in its size, its shape, and in the colour of its eyes; and the only essential difference consists in the small projecting feather which the Scops has on each side of its head.—But we shall describe these distinctions more minutely in the following articles.

Aldrovandus justly remarks, that most of the mistakes in Natural History arise from the confusion of names, and that the subject of Nocturnal Birds is involved in the obscurity and shades of night. What we have now mentioned will, I hope, in a great measure dispel the cloud; and to throw greater light, we shall subjoin a few remarks. The names *Ule, Eule* in German, *Owl, Owlet* in English, *Huette, Hulote* in French, are derived from the Latin *Ulula*, which imitates by its sound the cry of the large kind of Nocturnal Birds. It is probable, as Frisch remarks, that this appellation was first appropriated to the Great Earless Owl, but was afterwards applied to the small ones, from their resemblance in form and instinct; and at last became a general term, comprehending the whole genus. Hence proceeds that confusion which is but imperfectly remedied by annexing epithets that allude to their haunts, their shape, or their cry. For example *Stein-eule* in German, Stone-owl, is the *Chouette*, or the Brown Owl;

Kirch-

Kirch-eule in German, Church Owl, is the French *Effraie*, White Owl, which is alfo named *Schleyer-eule*, Winged Owl, *Perl-eule*, Pearl Owl. *Ohr-eule* in German, Horn Owl, is the *Hibou* of the French (Long-eared). *Knapp-eule*, Nutcracker Owl, is a name which might have been applied to all the large Owls, which make a noife like that action with their bills. *Bubo* in Latin, the *Eagle Owl*, is derived from *Bos*, from the refemblance of its note to the lowing of an ox. The Germans have imitated the found, *uhu.*

The three fpecies of Earlefs Owls, and the five fpecies of Eared Owls, which we have now diftinguifhed, include the whole genus of the Nocturnal Birds of Prey. They differ from the birds that commit their ravages in the day. 1. By the fenfe of fight, which is delicate, and unable to fupport the glare of light. The pupil contracts in the day-time, but in a manner different from that of cats; for it retains its form, and contracts equally in every direction, while that of cats becomes narrow and extended vertically. 2. By the fenfe of hearing, which appears to be fuperior to that of other birds, and perhaps to that of every other animal; for the drum of the ear is proportionally larger than in the quadrupeds, and befides they can open and fhut this organ at pleafure, a power poffeffed by no other animal. 3. By the bill, whofe bafe is not, as in thofe birds which prey in the day, covered with a thin naked fkin, but is fhaded with fea-

thers

thers projecting forward ; it is alfo fhort, and both mandibles are moveable like thofe of the parrokeets, which is the reafon that they fo often crack their bill, and can receive very large morfels, which their wide throat admits to be fwallowed. 4. By their claws, which have an anterior movable toe, that can be turned backwards at pleafure, and enables them to reft on a fingle foot more firmly and eafily than others.   5. By their mode of flying, which when they leave their hole, is a kind of tumbling, and is conftantly fideways, and without noife, as if they were wafted by the wind.—Such are the general diftinctions between the Nocturnal and Diurnal Birds of Prey : they have nothing fimilar but their arms, nothing common but their appetite for flefh and their inftinct for plunder.

The GREAT-EARED OWL*.

Le Duc, ou *Grand Duc*, Buff.
Strix Bubo, Linn. Gmel. Will. Kram. Briff. &c.
Ulula, Klein.
Bubo Maximus, Ger. Orn.
In Italian, *Gufo, Duco, Dugo*.
In Spanifh, *Bubo*.
In Portuguefe, *Mocho*.
In German, *Ubu, Huhu, Schuffut, Bhu, Becghu, Huhuy, Hub, Huo, Puhi*.
In Polifh, *Puhacz, Sowalezna*.
In Swedifh, *Uf*.

THE poets have confecrated the Eagle to Jupiter, and the Great-eared Owl to Juno. It is indeed the Eagle of the night, and the king of that tribe of birds which avoid the light of day, and refume their activity after the fhades of the evening defcend. At firft fight it appears as large and ftrong as the Common Eagle; but it is really much fmaller, and its proportions are quite different. The legs, the body, and the tail, are fhorter than in the Eagle; the head much larger; the wings not fo broad, and do not exceed five feet. It is eafily diftinguifhed by its coarfe figure, its enormous head,

* The Greek name Βυας is perhaps derived from Βυς, an *Ox*, from the refemblance of the cry of the Owl to the bellowing of an Ox. The Latin *Bubo* is faid to be formed from *Bufo*, a Toad, which it was fuppofed by the vulgar to breed. Does it not come from *Bos, Bovis*, an Ox, for the fame reafon as the Greek name?

the

THE EAGLE-OWL

the broad and deep cavities of its ears, the two
tufts which rife more than two inches and a half
on its crown ; its bill fhort, thick, and hooked ;
its eyes large, fteady, and tranfparent ; its pupils
large and black, furrounded with a circle of an
orange-colour ; its face encircled with hairs, or
rather fmall white ragged feathers, which terminate in the circumference of other fmall frizzled
feathers ; its claws black, very ftrong and hooked ; its neck very fhort ; its plumage of a rufty
brown, fpotted with black and yellow on the
back, and with yellow on the belly, mottled
with black fpots, and ribbed with a few bars of
a brown colour confufedly intermixed ; its feet
covered to the claws with a thick down and
rufty feathers * ; and laftly, its frightful cry
hiboo, hoohoo, boohoo, poohoo †, with which it in-
terrupts the filence of the night, when all the
other animals enjoy the fweets of repofe. It
awakens them to danger, difturbs them in their
retreat, purfues them, feizes them, or tears them
to pieces, and tranfports the fragments to the

* The female differs from the male in nothing except that its
plumage is generally darker.

† The following particulars are mentioned by Frifch, in regard
to the different cries of the Great-eared Owl, which he kept a long
time. " When it was hungry it uttered a found like that of the
word *Puhu*; if it heard an old perfon cough or hawk, it began very
loud, like the laughter of a peafant in liquor, and continued as long
as poffible without infpiring." " I fuppofe," fubjoins Frifch, " that
this was in the love feafon, and that it took the coughing for the
cry of its female : but when it cries through pain or fear, the found
is exceedingly ftrong and harfh, though a good deal like that of the
birds which prey by day."

caverns

caverns where it fixes its gloomy abode. It haunts only rocks, or old deferted towers that are fituated near mountains ; it feldom ventures into the plains ; it declines the boughs of trees, but commonly perches upon folitary churches and ancient caftles. Its prey confifts in general of young hares, rabbits, moles, and mice, which it fwallows entire, digefts the flefhy parts, and afterwards throws up the hair, bones, and fkin, rolled into a ball * ; it alfo devours bats, ferpents, lizards, toads, and frogs, and feeds its young with them. It is fo active in the breeding feafon, that its neft is quite crammed with provifions : it collects more than other birds of prey.

Thefe birds are fometimes kept on account of their fingular figure. The fpecies is not fo numerous in France as thofe of the other Owls ; and it is not certain whether they remain the whole year in the country. They however neftle fometimes in hollow trees, and oftener in

* " I have twice," fays Frifch, " had Great-eared Owls, and have kept them a long time. I fed them with ox-flefh and liver, of which they fwallowed very large bits. If mice were thrown to them they crufhed the bones with their bill, then fwallowed them one after another, fometimes to the number of five. After fome hours, the hair and bones are collected in the ftomach, and rejected through the bill. When they can get nothing elfe, they eat fmall and middle-fized river-fifh of every kind ; and after the bones are crufhed and rolled in the ftomach, they pufh thefe up the throat, and throw them out. They will not drink, a circumftance which I have obferved in fome diurnal birds of prey."—We may obferve that thefe birds can fubfift without drinking ; but they will often drink when they imagine themfelves to be concealed.

the

the crags of rocks, or in the holes of lofty old walls. Their neſt is near three feet in diameter, compoſed of ſmall branches of dry wood inter-woven with pliant roots, and ſtrewed with leaves. They commonly lay one or two eggs, and but ſeldom three ; theſe reſemble ſomewhat the colour of the plumage of the bird, and are larger than hens eggs. The young are very vo-racious ; and the parents are vigilant in provid-ing ſubſiſtence, which they procure in ſilence, and with much more agility than we ſhould ſuppoſe from their extreme corpulence. They often fight with the buzzards, are victorious in the combat, and ſeize the plunder. They ſup-port more eaſily the light of day than the other nocturnal birds ; for they leave their haunts earlier in the evening, and later in the morning. Sometimes the Great-eared Owl is ſeen attacked by flocks of crows, which accompany his flight and ſurround him by thouſands ; he withſtands their onſet*, drowns their hoarſe murmurs with his louder ſcreams, diſperſes them, and often when the light begins to fail he ſeizes ſome fated victim. Though his wings are ſhorter than thoſe of moſt of the birds that ſoar, he can riſe to a great height, eſpecially about twilight ; but at other times he generally flies low, and to ſhort diſtances. The Great-eared Owl is employed in falconry to attract the notice of the Kite, and

* Klein.

he is furnished with a fox-tail to heighten the
singularity of his figure. Thus equipped, he
skims along the surface of the ground, and
alights on the plain, without venturing to perch
upon a tree. The Kite perceives him from a
distance, and advances not to fight or attack
him, but to admire his odd appearance, and generally hovers about unguarded, till he is fur-
prised by the sportsman, or caught by the birds
of prey that are flown at him. Most of the
breeders of pheasants also keep a Great-eared
Owl, which they place in a cage among the
rushes in an open place, to draw together the
ravens and the crows, which gives them an
opportunity of shooting and killing a greater
number of these noisy birds, so alarming to the
young pheasants. To avoid scaring the pheasants, they shoot at the crows with a cross-bow.

With regard to the internal structure of this
bird, it has been remarked, that the tongue is
short and broad, the stomach capacious, the eye
inclosed in a cartilaginous coat in the form of a
capsule, the brain invested with a single coat
thicker than that of other birds ; and, as in the
quadrupeds, there are two membranes which
cover the *cerebellum*.

It appears that in this species there is a first
variety which includes a second; both are found
in Italy, and have been mentioned by Aldrovandus. The one may be called the *Black-*
winged

winged Great-eared Owl *; the fecond, the *Naked-footed Great-eared Owl* †. The firſt differs from the Common Great-eared Owl only by the colours of its plumage, which is browner or blacker on the wings, the back, and the tail. The ſecond, which reſembles it exactly in the deepneſs of its colour, is diſtinguiſhed by its legs and feet, which are but ſlightly ſhaded with feathers.

Beſides theſe two varieties, which are found in our own climate, there are others which occur in diſtant countries. The White Eagle-Owl of Lapland, mottled with black ſpots, and which is deſcribed by Linnæus, appears to be only a variety produced by the cold of the north ‡. Moſt of the quadrupeds are either white, or ſoon become ſo, within the polar circle, and many birds are ſubject to the ſame change. This bird, which is found in the mountains of Lapland, is white, ſpotted with black; and the difference of colour is what alone diſtinguiſhes it from the Common Great-eared Owl. We may therefore refer it to that ſpecies as a mere variety.

As this bird can bear both heat and cold, it is found in the north and ſouth of both continents;

---

* This is the firſt variety of the Great-eared Owl in the Linnæan ſyſtem, and the Athenian Horn-Owl of Edwards and Latham.

† This is the ſecond variety of Linnæus, and the Smooth-legged Owl of Latham.

‡ This is the *Strix Scandiaca,* a variety of the *Strix Virginiana* of Linnæus, or *Virginian Eared Owl.*

and

and not only is the species spread so extensively, but even the varieties. The *Jacurutu** of Brazil, described by Marcgrave, is exactly the same as our Common Great-eared Owl; and one brought from the Straits of Magellan differs little from the European species. That mentioned by the author of the Voyage to Hudson's-bay by the name of *Crowned Owl*†, and by Edwards *Eagle Owl* of Virginia, are varieties which occur in America the same with those in Europe; for the most remarkable difference between the Common Eagle-Owl, and that of Hudson's-bay and of Virginia ‡ is, that the tufts rise from the bill, and

* " It is equal in bulk to the geese; its head is round like that of a cat; the bill thick and black, the upper mandible projecting; the eyes large, prominent, round, and shining like crystal, within which a yellow circle appears near the margin; near the ears are feathers two inches long, which are erect, and tapering to a point like ears; the tail is broad, and the wings do not reach to its extremity; the legs are clothed with feathers as far as the feet, on which there are four toes, three before and one behind, each of which is furnished with a hooked nail that is black, more than an inch long, and very sharp; the feathers over the whole body are variegated elegantly with yellow, white, and blackish." MARCG.

† " The Great-crowned Owl is very common in the country about Hudson's-bay. It is a very singular bird, and its head is scarcely smaller than that of a cat; what are called its horns are feathers which *rise precisely above the bill*, where they are mixed with white, becoming by degrees of a brown red, spotted with black." *Voyage to Hudson's-bay.*

‡ " This bird," says Edwards, " is of the largest species of Owls, and much resembling in bulk the Horned-Owl, which we call the *Eagle-Owl*. Its head is as large as that of a cat. . . . The bill is black; the upper mandible is hooked, and projects beyond the lower, as in the Eagles; it is also sheathed with a skin in which the nostrils are placed, which is covered at the base with gray feathers that

and not from the ears.   But in the figures of
the three Eagle-Owls given by Aldrovandus,
the tufts rife from the ears in the firft only, or
the Common ; and in the others, which are but
varieties that occur in Italy, the tuft feathers are
not inferted at the ears, but at the bafe of the
bill, as in the Eagle-Owl of Virginia defcribed
by Edwards.   Klein was therefore rafh in afferting, that the Eagle-Owl of Virginia was a fpecies entirely different from that of Europe.   If
he had confulted the figures of Aldrovandus and
Edwards, he would have found that this diftinction, which only conftitutes a variety, occurs in

that encircle the bill ; the eyes are large, and the iris is fhining
and gold-coloured. . . . *The feathers which form thefe horns rife im-*
mediately above the bill, where they are mixed with a little white ;
but as they advance above the head they become of a brown red,
and terminate with black on the outfide ; the upper part of the
head, neck, back, wings, and tail, are of a dull brown, fpotted
and intermixed irregularly with fmall tranfverfe reddifh or cinereous lines. . . . The part of the throat under the bill is white ; fomewhat lower, orange-yellow, fpotted with black ; the lower part of
the breaft, the belly, the legs, and the under part of the tail, white
or pale gray, and pretty regularly croffed with brown bars ; the
infide of the wings is variegated, and coloured in the fame manner ;
the feet are covered as far as the nails with feathers of a whitifh
gray, and the nails are of a deep horn colour. . . . I drew this
bird after the life in London, whither it was brought from Virginia.   I have befide me another ftuffed one, which I received from
Hudfon's-bay ; it appears to me of the fame fpecies with the former, being of the fame fize, and differing only in the fhades of its
plumage."

I fhall obferve that there is only one character which feems to
imply that this bird is a permanent variety of the Great-eared Owl,
viz. that the tufts do not rife from the ears, but from the bafe of
the bill ; and as the fame bird is alfo found in Europe, we may regard it as compofing a diftinct family in this fpecies.

T 3

Italy

Italy as well as in Virginia, and that in general
the tufts of thefe birds do not proceed exactly
from the fide of the ears, but rather from below
the eyes, and the upper part of the bafe of the
bill. [A]

[A] This article includes two fpecies of Owls in the Linnæan
fyftem : —

Firft, *Strix Bubo*, or the Great-eared Owl, which has a rufous
plumage ; it is of the fize of the Eagle ; it inhabits Europe, and is
found, though rarely, in the north of England, Chefhire, and Wales.
It includes three varieties : firft, the Athenian Horn-Owl, which
is of a darker colour, and has more flender feet : fecondly, the
Smooth-legged Owl, whofe feet are naked : thirdly, the Magel-
lanic Eared Owl, or Jacurutu of Marcgrave, which is dufky-yel-
lowifh, variegated with white.

Secondly, *Strix Virginianus*, or the Virginian Eared Owl, the
Eagle Owl, or Horned Owl. " Its upper part is dufky, variegated
with delicate rufous and cinereous lines ; below, it is pale cinereous,
with dufky tranfverfe ftreaks ; the throat and fides of the breaft
ftreaked with dufky orange."   It is rather fmaller than the preceding ; it inhabits the north of Afia, and the whole of America,
where, during the night, it makes a hideous noife in the woods, not
unlike the hallooing of a man, and has often mifled travellers. The
Indians dread its ominous prefages, and are provoked at any perfon who mimicks its hooting.

THE LONG EARED OWL.

The LONG-EARED OWL*.

Le Hibou, ou *Moyen Duc*, Buff.
Strix Otus, Linn. Gmel. Will. Kram, &c.
Asio, Briss. and Klein.
Noctua Minor Aurita, Frisch.
Hornoder Ohr-eule, Gunth. Nest.
The Horn Owl, Albin.

THE ears of this bird are very wide, like those of the Great-eared Owl, and covered with a tuft formed of six feathers pointing forwards †; but these tufts are much shorter than those of the Great Owl, and hardly exceed an inch in length; they are however proportioned to its size, for it weighs only about ten ounces, and is not larger than a crow. It is therefore a species evidently different from that of the Great-eared Owl, which is about the bulk of a goose; and from that of the Scops, which is not larger than a blackbird, and in which the tufts above the ears are very short. I make this re-

---

* In Greek it was called Ωτος, from ας, the *ear*; in Latin, *Otis*, or *Asio*; in modern Italian it is termed *Gufo*, or *Barbagianni*; in Spanish, *Mochuelo*; in German, *Ohr-eule (Ear-owl)*, *Kautz-eule*, or *Käutzlein (Coot-owl)*; in Swedish, *Horn-ugla*; in Polish, *Cluknocny*, or *Sowa-ursata*.

† Aldrovandus says, that he observed that each of the feathers in this crest is susceptible of a separate motion, and that the skin which covers the cavity of the ears arises from the part of the inside next to the eye.

mark,

mark, becaufe fome naturalifts have confidered thefe as merely varieties of the fame fpecies. The Long-eared Owl meafures about a foot from the point of the bill to the claws; its wings extend three feet, and its tail is five or fix inches. The upper part of its head, neck, back, and wings, are marked with rays of gray, rufty, and brown; the breaft and belly are rufty, with irregular and narrow brown bars; the bill is fhort and blackifh; the eyes of a fine yellow; the feet covered with rufty-coloured feathers as far as the origin of the claws, which are pretty broad, and of a blackifh brown. We may alfo obferve, that the tongue is flefhy, and fomewhat forked, the nails very fharp, the outer claw moveable, and may be turned backwards; the ftomach capacious, the gall-bladder very large, the guts about twenty inches long, the two *cæcas* two inches and a half deep, and proportionally thicker than in the other birds of prey.

The fpecies is common, and much more nu- merous in our climate * than the Great-eared Owl, which feldom occurs in winter. The Long-eared Owl is ftationary during the whole year, and is even found more readily in winter than in fummer. It commonly lodges in old ruined buildings, in the caverns of rocks, in the hollows of aged trees, in mountain-forefts, and

* It is more common in France and Italy than in England. It is found very frequently in Burgundy, Champagne, Sologne, and in the mountains of Auvergne.

feldom

feldom ventures to defcend into the plains *.
When attacked by other birds, it makes a dex-
terous ufe of its talons and bill; and it even turns
on its back when its antagonift is too powerful.

It appears that this bird, which is common in
our part of Europe, is found alfo in Afia; for
Belon mentions his having met with it in the
plains of Cilicia.

This fpecies admits of feveral varieties, the
firft of which occurs in Italy, and has been defcribed by Aldrovandus. It is larger than the
common fort, and differs in the colour of its
plumage.

Thefe birds feldom take the trouble to conftruct a neft; for all the eggs and young which
I have received were found in the nefts of other
birds; often in thofe of magpies, which it
is well known conftruct a new one every year;
fometimes in thofe of buzzards; but I could
never procure a neft built by themfelves. They
generally lay four or five eggs, and the young,
which are at firft white, acquire their natural
colour in the courfe of fifteen days.

As this Owl can fupport cold, and is found in
Sweden † and in France, and paffes the winter
in our latitudes, it could migrate from one continent into the other. It appears that it is found

* " The Gufo (the Long-eared Owl) lodges in grottoes
and the hollows of trees, in the crannies and fiffures of walls, and
the roofs of uninhabited houfes, among precipices and in defert
tracts." *Olina Ucceller.*

† Linn. *Faun. Suec.*

in

in Canada *, and in many other parts of North America ; and probably the Owl of Carolina defcribed by Catefby, and that of South America mentioned by Father Feuillée †, are only varieties of our fpecies, occafioned by the difference of climates ; efpecially as they are nearly of the fame fize, and differ only in the fhades and diftribution of their colours.

The Long-eared and Tawny Owls ‡ are employed to attract the birds by their call ; and it is obferved that the large birds more readily obey

* The following paffage from Charlevoix muft refer to the Long-eared Owl :—" There is heard almoft every night in our ifland a kind of Tawny Owl which they call *Canot,* and which utters a mournful cry, as if it hallooed *au canot* (to the canoe), whence its name. Thefe birds are not larger than turtles, but they are exactly fimilar in their plumage to the Long eared Owls that are common in France. They have two or three fmall feathers on both fides of the head, which look like ears. Sometimes feven or eight of them affemble on our houfe-tops, and fcream without interruption the whole night." The fize here indicated would lead to fuppofe that this bird is the Scops ; but the projecting feathers of the head fhew that it is a variety of the Long-eared Owl.—The fame author adds, that the Tawny Owl of Canada differs from that of France in nothing but in having a peculiar cry, and being furnifhed with a little white ruff about its neck.

† *Bubo ocro-cinereus Pectore Maculofo,* Feuillée, *i. e.* " The Afh-coloured Ferruginous-Owl, with a fpotted breaft." The *Tecolotl,* which is found in Mexico and New Spain, is perhaps the fame bird ; though this is only conjecture ; for Fernandez has given no figure, or full defcription.

‡ " The Gufo, or Great Nocturnal Owl, is of the fhape of the Tawny Owl, and about the bulk of a hen, with feathers on the fides of the head that appear like two horns of a yellow colour, and mixed with a border of black. It ferves as a call for the large birds, fuch as all the crows and kites, as the Tawny Owl for every kind of fmall birds." OLINA.

the

the note of the Long-eared Owl, which is a kind
of plaintive cry or hollow moaning, *clow̄, cloūd*,
inceſſantly repeated during the night ; and that
the ſmall birds reſort in greater numbers to the
invitation of the Tawny Owl, which is louder
and a kind of hallooing, *hŏhō, hŏhō*. Both
theſe in the day-time make ludicrous geſtures in
the preſence of men, and other animals.   Ariſtotle aſcribes this buffoonery to the Long-eared
Owl alone, *Otus*. Pliny beſtows it on the Scops,
and terms it *Motus Satyricos* (i. e. *Satiric Movements*).   But the Scops of Pliny is the ſame
with the *Otus* of Ariſtotle ; for the Latins confounded theſe names together, and united them
into one ſpecies, qualifying them only by the
epithets of *great* and *ſmall*.

What the ancients have related with reſpect
to theſe awkward motions and ridiculous geſtures,
muſt be principally applied to the Long-eared
Owl ; and, as ſome philoſophers and naturaliſts
have pretended that this was not an Owl, but
quite a different bird, which they term the *Lady
of Numidia*, I ſhall here diſcuſs the queſtion,
and remove the miſtake.

The Anatomiſts of the Academy of Sciences
are thoſe to whom I allude ; who, in their deſcription of the Lady of Numidia *(Demoiſelle
de Numidie)*, have endeavoured to eſtabliſh this
opinion, and ſtate their reaſons in the following
terms :—" The bird," ſay they, " which we de-
" ſcribe, is called the *Lady of Numidia ;* becauſe
" it

" it is a native of that part of Africa, and feems
" to imitate in fome degree the light air and
" fkip of a lady who affects a graceful motion.
" It is more than two thoufand years fince na-
" turalifts have fpoken of this bird, and remark-
" ed this peculiarity of inftinct.   Ariftotle gives
" it the name of *Juggler*, *Dancer*, and *Buffoon*,
" aping what it fees.—It would feem this danc-
" ing mimicking bird was rare among the an-
" cients, becaufe Pliny believes that it was fa-
" bulous, referring this animal, which he calls
" *Satyricus*, to the clafs of Pegafufes, Griffons,
" and Syrens. It has probably been till now un-
" known to the moderns, fince they do not de-
" fcribe it from their own obfervation, but only
" from the writings of antiquity, in which it re-
" ceived the names of *Scops* and *Otus* from the
" Greeks, and *Afio* from the Latins, and which
" they had termed *Dancer*, *Juggler*, and *Comedi-*
" *an;* fo that it muft be inquired, whether our
" Lady of Numidia is really the *Scops* or *Otus* of
" the ancients. The defcription which they have
" given us confifts of three particulars :—1ft, It
" imitates geftures.—2d, It has tufts of fea-
" thers on both fides of the head, like ears.
" —3d, The plumage is, according to Alexander
" the Myndian, in Athenæus, of a leaden colour.
" But all thefe properties belong to the *Lady of*
" *Numidia;* and Ariftotle feems to mark their
" manner of dancing, which is that of leaping
" the one before the other, when he fays, that
" they

" they are caught when they dance one oppofite
" to the other. Belon however believes, that
" the *Otus* of Ariftotle is the Owl, for this
" only reafon, that that bird makes many gef-
" tures with its head: moft of the tranflators of
" Ariftotle, who are alfo of our opinion, found
" it upon the name *Otus*, which fignifies having
" ears; but they are not peculiar to the Long-
" eared Owl; and Ariftotle evidently fignifies that
" the *Otus* is not the Long-eared Owl, when he
" fays, that the *Otus* refembles it; and this re-
" femblance is probably not in regard to the
" ears. All the *Ladies of Numidia* which we
" have diffected, had on the fides of the ears
" thefe feathers, which have given occafion to
" the name *Otus* of the ancients. . . . Their
" plumage was of an afh-colour, fuch as
" is defcribed by Alexander the Myndian as be-
" longing to the *Otus*."

Let us compaare Ariftotle's defcription of the
Otus with that of the Academicians. " The
" *Otus* is like the Owl, being furnifhed with
" fmall projecting feathers about the ears, whence
" its name, *Otus* or Eared; fome call it *Ullula*,
" others *Afio*; it is a babbler, a tumbler, and a
" mimic, for it imitates dancers. It is caught like
" the Owl, the one bird-catcher going round it,
" while it is intent upon the other."

The *Otus*, that is, the Long-eared Owl, is like
the *Noctua* or Tawny Owl; they refemble in

fact

fact, in their size, their plumage and natural habits; both are nocturnal birds of contiguous species; whereas the *Lady of Numidia* is six times thicker and larger, and is of a quite different shape, and of a different genus, being by no means a nocturnal bird. The *Otus* differs from the *Noctua* only by the tufts on the head near the ears, and Aristotle remarks this distinction. These are small feathers, straight and tufted, not the long ones that fall back, and hang from each side of the head, as in the *Lady of Numidia*. We cannot therefore refer the word *Otus* to this bird, but evidently to this Long-eared Owl (*Noctua Aurita*); and this inference is confirmed by what Aristotle immediately adds, " some call it *Ulula*, and others *Asio*." Nothing therefore is more groundless, in my opinion, than the pretended resemblance which they have endeavoured to trace between the *Otus* of the ancients and the *Lady of Numidia*, the whole o which is founded on some ludicrous gestures and motions which distinguish this sprightly bird; but the Long-eared Owl is still entitled to the epithets *Screamer* *, *Mimic*, *Buffoon*. The other character which Aristotle mentions, that this bird is easily caught, as the other Owls, can only be applied to this bird of night. . . . I might enlarge

* Frisch, speaking of this Owl, says, that its cry is very frequent and strong; and he compares it to the hooting of children running to make game of one; but that this cry is common to several kinds of Owls.

upon

upon this fubject, and bring paffages from Pliny
to fupport my criticifm; but a fingle remark
will remove all doubt. The ludicrous geftures
afcribed by the ancients to the Long-eared Owl,
belong to moft of the nocturnal birds*; their
afpect is marked with aftonifhment, they turn
their neck frequently, move their head upwards,
downwards, fideways, crack their bill, tremble
with their legs, fhifting their toe backwards and
forwards. thefe geftures may be obferved in birds
kept in captivity; but unlefs they are caught
while young we cannot rear them; for thofe
grown up, obftinately refufe all fuftenance. [A]

* All the Owls can turn their head like the *Wryneck*. If fomething uncommon occur, they open their large eyes, briftle their
feathers, and look twice as big; they alfo fpread their wings,
cower, or fquat, but fuddenly rife again as if aftonifhed; and twice
or thrice they crack their bills. *Id.*

[A] The fpecific character of the *Long-eared Owl, Strix Otus,*
LINN. is, " that the tufts of its ears confift of fix feathers." It
inhabits Europe, America, and the north of Afia, and is even found
in the warm climate of Egypt. It lives in the woods remote from
the fea, near the fettlement of Hudfon's-bay. It approaches the
dwellings, and is very noify. It builds in the trees, and never
migrates.

The SCOPS-EARED OWL*.

Le Scops, ou *Petit Duc*, Buff.
Strix Scops, Linn. Gmel. Will. Briff. Klein.
Hornoder Ohr-eule, Gunth. Neft.
Chiu, Alloccarello, Chivino, Zinn.

THIS is the third and laft fpecies of the
Eared-Owls. It is eafily diftinguifhed
from the other two; for its fize exceeds not that
of the black-bird, and the tufts over the ears
projeƐt only half an inch, and confift of a fingle
feather †:—alfo, its head is much fmaller in proportion to its body, and its plumage is more elegantly and diftinƐtly mottled, being variegated
with grey, ferruginous brown, and black, and
its legs are clothed to the origin of the nails with
feathers of a rufty grey mixed with brown fpots.
It is diftinguifhed alfo by its inftinƐt; for in
fpring and autumn it migrates into other climates.
It feldom paffes the winter in our provinces, but
departs after, and returns a little before, the fwallows. Though the Scops-eared Owls prefer the
high grounds, they crowd where field-mice

* From the Greek Σκωψ, which feems to be formed of Σκια, a
fhadow, and Ωψ, the face; probably becaufe it avoids the light.
† " The ears, or fmall feathers that projeƐt like ears, fcarcely
appear in the dead fubjeƐt; they are more obvious in the living
animal, and confift of only a fingle featherlet." ALDROV.

abound,

THE SCOPS OWL .

abound, and are ufeful in extirpating thefe de-
ftructive animals, which, in fome years, multiply extremely, and confume the grain, and deftroy the roots of plants that are the moft neceffary to the fupport of man. It has been often
obferved, that when this calamity is threatened,
the Scops affemble in flocks, and make war fo
fuccefsfully againft the mice, that in a few days
they entirely clear the field *. The Long-eared
Owls alfo gather fometimes to the number of an
hundred : and of this fact we have been twice
informed by eye-witneffes ; but it feldom occurs.
Perhaps thefe affemblies are formed with the
view of beginning a diftant journey : it is even
probable, that they migrate from the one continent into the other : for the Bird of New Spain,
mentioned by Nierenberg by the name of *Talchicualti*, is either of the fame fpecies, or of one
nearly allied to that of the Scops †. But though
it travels in numerous flocks, it is ftill rare and

---

* Dale, in his appendix to his Hiftory of Harwich, quotes two
inftances of this from Childrey. " In the year 1580, at Hallow-
tide, an army of mice fo overrun the marfhes near South-Minfter,
that they eat up the grafs to the very roots. But at length a
great number of *ftrange painted Owls* came and devoured all the
mice. The like happened in Effex anno 1648." Dale afcribes
this to the Long-eared Owl, yet the appellation of *ftrange painted
Owls* feems to mark the Scops.

† " The *Talchicualti* feems to be a foreign kind of *otus*; it
is horned or eared, the body fmall, the bill fhort and turned up,
the pupil black, the iris yellow black-colour, clothed with dufky
and cinereous feathers as far as the legs, which are black and incurvated at the nails. In other refpects, it is like our *otus*."

not eafily caught; nor have I been able to procure either the eggs or the young; it was even
difficult to inftruct the fportfmen to diftinguifh it
from the Little Owl, becaufe both thefe birds
are of the fame fize, and the fhort prominent
feathers which form the fpecific character of the
Scops, cannot be perceived at a diftance.

The colour varies much, according to the age,
the climate, and perhaps the fex; they are all
gray when young, but as they grow up, fome
are browner than others; the colour of the eyes
feems to correfpond to that of the plumage;
thofe that are gray have eyes of a pale yellow, in
others the colour is deeper; but thefe differences
are flight, and ought not to alter the claffification.

The ALUCO OWL.

La Hulotte, Buff.
Strix Aluco, Linn. Gmel. and Scop.
Ulula, Briff. and Will.
Noctua Major, Frif.
Ulula Vulturina, Klein.
Black Owl, Albin.
Brown Owl *, Penn. and Lewin.

THE Aluco, which may be alfo named the
Black Owl, and which the Greeks called
the *Nycticorax* or *Night Raven,* is the largeft of
all the Owls. It is near fifteen inches long from
the point of the bill to the claws; its head is
large, round, and without tufts ; its face funk as
it were in the plumage; its eyes are buried in
greyifh ragged feathers ; the iris blackifh, or
rather deep brown ; the beak of a yellow or
greenifh white ; the upper part of the body a
deep iron-gray, mottled with black and whitifh
fpots; the under white, with longitudinal and
tranfverfe bars ; the tail fomewhat more than
fix inches, the wings ftretching a little beyond
its extremity, and when fpread, meafure three
feet ; the legs covered to the origin of the nails
with white feathers, fprinkled with black

---

* In Greek Νυκλικοραξ ; in Latin, *Ulula,* from its howling cry;
in Italian, *Alocho;* in German, *Huhu* ; in Polifh, *Lelok, Sowka,* and
Pufzxik ; in Portuguefe, *Corufa.*

points:

points *: thefe chara&ers are fully fufficient to
diftinguifh the Aluco from all the others; it flies
lightly, and not ruftling with its wing, and always fideways, like the reft of the Owls. Its
cry †, *hoō, ōō, ōō, ōō, ŏŏ, ŏŏ, ŏŏ*, refembling the
howling of ~~wolves~~ *(ululare)*, was the foundation of its name *ulula* among the Romans; and
the fame analogy has led the Germans to apply
the term *hoō, hoō*.

The Aluco lodges during fummer in the woods,
and conftantly in hollow trees. Sometimes it
ventures in winter to approach our habitations;
it purfues and catches fmall birds; but field-mice
are its more ufual prey; it fwallows them whole,
and afterwards difcharges by its bill the fkins
rolled into balls. When it is unfuccefsful in the
field, it reforts to the farm yards and barns, in
queft of mice and rats. It retires early in the
morning to the woods, about the time that the
hares return to their retreats, and conceals itfelf
in the thickeft copfe, or remains the whole day
motionlefs in the foliage of the fhadieft trees.
During inclement weather, it lodges in hollow
trees in the day, and makes its excurfions in the

---

* We may add a diftinguifhing mark, that the uttermoft feather of
the wing is two or three inches fhorter than the fecond, and this alfo
an inch fhorter than the third, and that the longeft of all are the
fourth and fifth; whereas in the White Owl, the third one is the
largeft of all, and the uttermoft is only half an inch fhorter.

† "This bird howls in the night, efpecially when it freezes,
with fo gloomy a moan, as to terrify women and children." SA-
LERNE.

night.

night.—Thefe inftinctive habits are common to it and to the Long-eared Owl, as well as that of depofiting its eggs in the nefts of other birds, fuch as the Buzzards, the Keftrels, the Crows, and Magpies.   It generally lays four eggs, of a dirty gray colour, round fhaped, and nearly as large as thofe of a fmall pullet. [A]

[A] The fpecific character of the Aluco is, that "its head is fmooth, its body ferruginous, its irides black."   It is a native of Europe, and is held facred among the Calmuck Tartars.

The TAWNY OWL.

Le Chat-Huant *, Buff.
Strix Stridula, Linn. Gmel. Brun. and Kram.
Noctua Major, Frif.
Strige, Zinn.
The Common Brown or Ivy Owl †, Will. and Alb.

AFTER the Aluco, diftinguifhed from the reft of the Earlefs Owls by its magnitude and its blackifh eyes, come the Tawny with bluifh eyes, and the White with yellow eyes. They are both nearly of the fame fize; being thirteen inches long from the point of the bill to the claws; fo that they are only two inches fhorter than the Aluco, but appear proportionally more flender.—The Tawny Owl is diftinguifhed by its bluifh eyes, the beauty and variety of the colours of its plumage, and its cry *hŏbō, hŏbō, hŏbŏhŏbŏ*, by which it feems to fhout or halloo with a loud voice.

Gefner, Aldrovandus, and many other naturalifts after them, have ufed the word *Strix* to diftinguifh this fpecies; but I believe that they

---

* i. e. The Hooting Cat.

† In Greek, Γλαυξ, from γλαυκος, fea-green, on account of its colour; in Latin, *Noctua*, being a nocturnal bird; in German, it is termed *Milch Sanger* (Milk Singer), *Kinder* (the Child), *Melcker* (the Milker), and *Stock-Eule* (the Stick Owl).

are

THE TAWNY OWL.

are miftaken, and that the term ought only to
be applied to the White Owl. *Strix* taken in
this fenfe, as denoting a bird of night, muft be
confidered as rather a Latin than a Greek word.
Ovid gives its etymology, and marks with fufficient precifion to which of the Nocturnal
Birds it ought to be referred, in the following
paffage :

——————Strigum
Grande caput, ftantes oculi, roftra apta rapinæ,
   Canities pennis, unguibus hamus ineft.
Eft illis ftrigibus nomen ; fed nominis hujus
   Caufa quod horrenda ftridere nocte folent *.

A large head, fixed eyes, a bill fitted for rapine, hooked nails, are characters common to
all thefe birds ; but the whitenefs of the plumage,
canities pennis, belongs more properly to the
White Owl than to any other ; but what in my
opinion decides the queftion is, that the word
ftridor, which in Latin expreffes a grating noife
refembling that of a faw, marks precifely the cry
grĕ, grëi of the White Owl ; whereas the cry of
the Tawny Owl is rather a loud hallooing than a
creeking noife.

---

* Thus tranflated by Maffey—
   " Large is their head, and motionlefs their eye,
Hook-billed, fharp-clawed, and in the dufk they fly.
      *   *   *
Screech-Owls they're called ; becaufe with difmal cry,
In the dark night, from place to place they fly."
                 Trifti. lib. vi. *fub init.*

The

The Tawny Owls are fcarcely found any where but in the woods. In Burgundy they are more common than the Alucos; they lodge in hollow trees, and I have received fome in the fevereft winters; which fact feems to prove that they are ftationary in the country, and feldom approach the habitations of men. Frifch gives the Tawny Owl as a variety of the fpecies of Aluco, and takes the male for another variety; but if we admit this claffification, we muft deftroy invariable characters, which feem to be numerous and diftinct.

The Tawny Owl is found in Sweden and other northern countries, and hence it has migrated into the continent, or is found in America, even between the tropics. There is in the cabinet of Mauduit a Tawny Owl, which he received from St. Domingo, and which feems to be a variety of the European fpecies, differing only by the uniformity of the colours of its breaft and belly, which are ferruginous, and almoft fpotlefs, and by the deeper fhades of the upper part of the body. [A]

[A] The fpecific character of the Tawny Owl, or *Strix Stridula*, Linn. is,—" Its head is fmooth, its body ferruginous, and the third feather of the wing longer than the reft." It inhabits the more fouthern deferts of Europe and Tartary; and even in England, it is pretty frequent in the woods. It breeds in the rook's nefts. It is not found in Siberia. Weighs nineteen ounces.

THE WHITE OWL.

The WHITE OWL.

L'Effraie, ou *La Fresaie,* Buff.
Strix Flammea, Linn. Gmel. Mull. and Bor.
Aluco, Briff. Ray, Brun. and Klein.
Aluco Minor of Aldrovandus, Will.
Tuidara, Marcgr.
Schläffer Eule, Perle-Eule , Frifch.

THE White Owl alarms the timid by its
blowing notes, *fhē, fhēï, fhēïē* ; its fharp doleful cries, *grēï, grĕ, crĕï,* and its broken accents
which often difturb the dread filence of night.
It is in fome degree domeftic; it inhabits the
moft populous towns, towers, belfries, the roofs
of churches, and other lofty buildings, which
afford it retreat during the day. It leaves its
haunts about twilight, and continually repeats
its blowing, which refembles the fnoring of a
man which fleeps with his mouth open. When
it flies or alights, it utters alfo different fharp
notes, which are all fo difagreeable, that, joined
to the awfulnefs of the fcene, re-echoed from
the tombs and the churches in the ftillnefs and

---

* The Greek name Ελεος is perhaps taken from the fame word,
which fignifies *pity,* alluding to its mournful cries; the German
appellations allude to its figure and manner of living; *Schleyer-
Eule,* Veiled Owl; *Perle-Eule,* Pearled Owl; *Kirch-Eule,* Church
Owl; and *Schlaffer Eule,* the Sleepy Owl.

dark-

darkneſs of night, inſpire dread and terror in
the minds of women and children, and even of
men who are under the influence of the ſame
prejudices, and who believe in omens and
witches, in ghoſts and apparitions. They regard the White Owl as a funereal bird, and the
meſſenger of death; and they are impreſſed
with an idea, that if it perches upon a houſe,
and utters cries a little different from ordinary, it then ſummons the inhabitant to the
tomb.

It is readily diſtinguiſhed from the other
Earleſs Owls, by the beauty of its plumage; it
is nearly of the ſame ſize with the Tawny Owl,
ſmaller than the Aluco, and larger than the
Brown Owl, of which we ſhall treat in the following article. Its extreme length is a foot, or
thirteen inches; its tail meaſures only five
inches; the upper part of its body is yellow,
waved with gray and brown, and ſprinkled with
white points; the under part white, marked
with black ſpots; the eyes regularly encircled
with white feathers, ſo ſlender that they might
be taken for hairs; the iris is of a fine yellow,
the bill white, except the end of the hook, which
is brown; the legs covered with white down,
the claws white, and the nails blackiſh. There
are others which, though of the ſame ſpecies,
ſeem at firſt to be very different; in ſome the
breaſt and belly are of a fine yellow, ſprinkled
with the ſame black points; in others they are

perfectly

perfectly white; in others they are yellow, and without a fingle fpot.

I have had feveral alive. They are eafily caught, by placing a fmall net at the holes where they lodge in old buildings. They live ten or twelve days in the cages where they are fhut, but they reject all fuftenance, and die of hunger. They continue motionlefs during the day, but mount the top of the rooft in the night, and whiftle the note *fhē, fhēi*, by which they feem to invite the others; and indeed I have often feen them attracted by the calls of the prifoner, alight near the cage, make the fame whiftling noife, and allow themfelves to be caught in the net. I never heard them when confined utter the grating cry *(ftride) crĕī, grĕī*; this found is given only in the flight, when they are in perfect freedom. The female is fomewhat larger than the male, and the colours of its plumage are lighter and more diftinct; and of all the nocturnal birds its plumage is the moft beautifully varied.

The fpecies of the White Owl is numerous, and very common in every part of Europe. It is alfo found through the whole extent of the continent of America. Marcgrave found it in Brafil, where the inhabitants call it *Tuidara*.

The White Owl does not, like the Aluco and the Tawny Owls, depofit its eggs in the nefts of other birds. It drops them in the bare holes of walls, or in the joifts of houfes, and alfo in the hollows of trees; nor does it fpread roots or

leaves

leaves to receive them. It begins early in the
fpring, in the end of March, or the beginning
of April. It lays five egs, fometimes fix or feven,
of a longifh fhape, and whitifh colour; it feeds
its young with infects and fragments of mice.
They are white at firft, and are not an unpleafant meal at the end of three weeks, for they
are fat and plump. Their parents clear the
churches of the mice; frequently drink or rather eat the oil from the lamps, efpecially when
it has congealed; fwallow mice and fmall birds
whole, vomiting afterwards the bones, feathers,
and fkin. Their excrements are white and
liquid like thofe of the other birds of prey. In
fine weather, moft of thefe birds vifit the neigh-
bouring woods in the night, but return to their
ufual haunts in the morning, and there flumber
and fnore till dark, when they fally from their
holes, and fly tumbling almoft to the-ground.
In the fevere feafons five or fix are fometimes
difcovered in the fame hole, or concealed in the
fodder, where they find fhelter, warmth, and
food; for the mice are more plentiful then in the
barns than at any other time. In autumn they
often pay a nightly vifit to the places where the
fprings are laid for the wood-cocks and thrufhes;
they kill the wood-cocks, which they find hanging, and eat them on the fpot; but they fometimes carry off the thrufhes and other fmall birds
that are caught, often fwallowing them entire with their feathers, but generally when
they

they are larger, plucking them previoufly.—
Thefe inftincts, and that of flying fideways with
ruftling wings, are common to the White, the
Aluco, and the Tawny Owls. [A]

[A] The fpecific character of the *White Owl, Strix Flammea,*
Linn. is, that " its head is fmooth, its body yellowifh, with
white points ; below it is whitifh, with blackifh points." It is common in England.   It is found through Europe and America, but
not farther north than the latitude of Sweden.   In Tartary it receives divine honours, from a tradition that it was inftrumental in
faving the Emperor Zingis Khan ; and even at prefent, the Kal-
mucks have retained the cuftom of wearing a plume of its feathers
on great feftivals.

The BROWN OWL*.

Le Chouette, ou *La Grande Chevêche*, Buff.
Strix Ulula, Linn. Gmel. Mull. and Georgi.
Ulula Flammeata, Frisch.
Strix Cinerea, Ray, Will. and Browsk.
Noctua Major, Briss.
Noctua Saxatilis, Gesn. and Aldrov.
Grey Owl, Will.
Great Brown Owl, Alb.

THIS species is pretty common, but does not
frequent our habitations so much as the
White Owl. It haunts quarries, rocks, ruins,
and deserted edifices; it even prefers mountainous tracts, craggy precipices, and sequestered
spots; but it never resorts to the woods, or
lodges in hollow trees. The colour of its eyes,
which is a bright yellow, distinguishes it from
the Aluco and the Tawny Owls. The difference is more slight between it and the White
Owl; because in both, the iris is yellow, sur-
rounded with a circle of small white feathers;
the under-part of the belly is tinged with yellow; and their size is nearly the same. But the
Brown Owl is of a deeper colour, marked with
larger spots resembling small flames; whereas

---

* Perhaps its Greek name Αιγωλιος is from Αιξ, αιγος, a *goat,*
because like that animal it is fond of rocks.—The appellation in
German is *Stein-Eule,* for the same reason. In Polish, it is called
Sowa.

thofe

THE BROWN OWL.

thofe of the White Owl are only little points
or drops; hence the former has been termed
Noctua Flammeata, and the latter *Noctua Guttata*.
The feet of the Brown Owl are clofely covered
with feathers, and the bill is brown; while the
bill of the White Owl is whitifh, and brown
only near the tips. In this fpecies alfo, the plumage of the female is marked with fmaller fpots
than the male, and its colours are more dilute.
Belon confiders the White Owl as allied to the
Little Owl; and indeed they bear a refemblance
in their figure and inftincts; and in German
they both have the generic name *Kautz (Coot)*.
Salerne informs us, that in the province of Orleans the labourers have a great efteem for this
bird, becaufe it deftroys the field-mice; that in
the month of April it utters day and night the
found *goo* in a foft tone; but before rain it
changes this note into *goyong*; that it builds no
neft, and lays only three eggs, which are entirely white, perfectly round, and about the fize
of thofe of a wood-pigeon. He adds, that it
lodges in hollow trees, and that Olina was
grofsly miftaken when he afferted that it hatches
in the two laft months of winter. The laft circumftance, however, is not far from the truth;
for this bird, and thofe of the fame kind, lay
their eggs in March, and the incubation muft
take place about the fame time. Nor is it
caught in hollow trees, but, as we have already
faid, it haunts the rocks and caverns. It is considerably

fiderably fmaller than the Aluco, and even than
the Tawny Owl, its extreme length being only
eleven inches.

It appears that this Brown Owl which is common in Europe, efpecially in the hilly countries,
is alfo found in the mountains of Chili; and
that the fpecies defcribed by Father Feuillée by
the epithet of *Rabbit*, becaufe it was difcovered
in a hole in the ground, is only a variety of the
European kind, differing by the diftribution of
its colours. If indeed it had made the exca-
vation itfelf, as Father Feuillée imagines, we
muft confider it as entirely diftinct from any
Owl even of the ancient continent *. But fuch
a fuppofition is unneceffary; it is moft likely
that, guided by inftinct, it only crept into holes
which it found already formed. [A]

* Father du Tertre, fpeaking of a nocturnal bird called the *devil*
in our American iflands, fays, that it is as large as a duck; that its
afpect is hideous; its plumage mixed with white and black; and
that it lives on the higheft mountains; that it *burrows* like a rabbit in the holes which it makes in the ground, where it lays its eggs,
hatches, and raifes its young. . . . that it never defcends from the
mountains, except in the night; and, when it is on the wing, it
utters a melancholy frightful cry.—This is certainly the fame bird
with the one mentioned by Feuillée, and with the Brown Owl.

[A] The fpecific character of the Brown Owl, or *Strix Ulula*,
given by Linnæus, is, " That the upper part of its body is dufky,
with white fpots; the tail-feathers infcribed with white lines."—
The defcription of Latham is more accurate and complete : " Its
head is fmooth; the upper part of the body is tawny, with dufky
longitudinal fpots; below whitifh with dufky lines; the tail marked
with dufky bars." It is fifteen inches and a half long, and weighs
fourteen ounces. It is not common in England. It includes two
varie-

varieties : 1. *The Arctic Owl, Strix Artica,* of which the body is
ferruginous above, with black fpots ; and the orbits, the bill, and
a bar under the wings, are black.   It inhabits the northern parts
of Sweden.   It is eighteen inches long.—2. *The Caſpian Owl,
Strix Accipitrina,* of which the upper part of the body is ſlightly
yellowiſh ; and below it is yellowiſh white, with blackiſh longitudinal ſpots.   It inhabits the Caſpian Sea, the ſouthern parts of
Ruſſia and Tartary, and occupies deſerted neſts,

The LITTLE OWL*.

La Chevêche, ou *Petite Chouette,* Buff.
Strix Passerina, Linn. Gmel. Scop. Brun. Mul. Kram, &c.
Noctua Minor, Ray, Will. and Klein.
La Civetta, Olin. and Zinn.

THE Little Owl and the Scops Owl are nearly of the same size, both being the smallest of the genus. They are seven or eight inches long from the point of the bill to the claws, and not larger than a blackbird; but they are still a distinct species; for the Scops is furnished with very short slender tufts, consisting of a single feather on each side of the head, which are entirely wanting in the Little Owl: besides, the iris is of a paler yellow, the bill brown at the base, and yellow near the point; but that of the Scops is entirely black. It may be readily distinguished by the difference of colours, by the regular disposition of the white spots on the wings and the body, by the shortness of its tail and wings, and by its ordinary cry, *poŏpoŏ, poŏpoŏ,* which it constantly reiterates

* The Greeks and Romans seem to have had no name appropriated to this species; and probably they confounded it with the Scops Owl, or *Asio.* This is the case in the modern languages: both are termed *Zuetta* or *Civetta* in Italian; *Sechuza* in Spanish; *Mocho* in Portuguese; *Kautzlein* in German; and *Szowa* in Swedish.

while

THE LITTLE OWL.

while it flies ; and another note which it has
when fitting, and which refembles the voice of a
young man, who repeatedly calls *aīmĕ, hēmĕ, efmĕ* *.
It feldom haunts the woods ; but its ordinary
abode is among folitary ruins, caverns, and old
deferted buildings, and it never lodges in hollow trees. In all thefe refpects it refembles moft
the Brown Owl. Nor is it entirely a bird of
night ; but fees much better in the day than the
other nocturnal birds, and even chaces the fwallows and other fmall birds, though with very
little fuccefs. It is more fortunate in the fearch
for mice, which it fwallows, not entire, but tears
them in pieces with its bill and claws ; and it
even plucks the birds neatly before it eats them ;
and in this inftinct it differs from the other Owls.
It lays five eggs, which are fpotted with white
and yellow, and conftructs its rude, and almoft
bare neft in the holes of rocks, and old walls.
Frifch obferves, that this bird loves folitude, and
haunts churches, vaults, and cemeteries, the refidence of the dead ; that it is fometimes called
Church-Owl, Corpfe-Owl; and that as it has been
remarked to flutter about houfes where there were
perfons dying, the fuperftitious people name it *the*

---

* Happening to fleep in one of the old turrets in the caftle of
Montbard, a Little Owl alighted on the window-frame, and before
day-break, at three o'clock in the morning, awakened me with its
cry, *hēmĕ, ĕdmĕ*. As I was liftening to this found, which was the
more remarkable as it was clofe befide me, I heard one of my fervants who flept in the room over mine open the window, and deceived by the refemblance of the fcream *ĕdmĕ*, call out, *Who's there
below ? my name is not Edme, it is Peter.*

bird of death, and imagine that it portends approaching diffolution. Frifch does not reflect that thefe gloomy images are connected only with the White Owl, and that the Little Owl is much more rare; that it hovers not about churches, nor has the plaintive moan or the piercing intimidating cry of the other. At any rate, if the Little Owl be reckoned *the bird of death* in Germany, it is the White Owl that is held ominous in France. The Little Owl which Frifch has figured, and which occurs in Germany, appears to be a variety of ours: its plumage is much darker, and its iris black. There is alfo a variety in the king's cabinet, which was fent from St. Domingo, and which is not fo white on the throat, and whofe breaft and belly are regularly marked with brown tranfverfe bars; while, in our Little Owl, the brown fpots are fcattered confufedly.

It may be proper to prefent a clear concife view of the diftinguifhing characters of the five fpecies of Earlefs Owls, of which we have treated. 1. The Aluco is the largeft; its eyes are black; it may be termed *The Large Black Earlefs Owl with Black Eyes*. 2. The Tawny Owl is much fmaller than the Aluco; its eyes blueifh; its plumage rufty, tinged with iron-grey; the bill greenifh white; and may be named *The Rufty and Iron-grey Earlefs Owl*

Owl with Blue Eyes. 3. The White Owl is nearly of the fame fize with the Tawny; its eyes yellow; its plumage whitifh yellow, variegated with very diftinct fpots; the bill white, and the end of the hook brown; and may be called *The White or Yellow Earlefs Owl with Orange-Eyes.* 4. The Brown Owl is not fo large as the Tawny or White, but nearly as thick; its plumage brown; its eyes of a fine yellow; its bill brown; and may be termed *The Brown Earlefs Owl, with Yellow Eyes and a Brown Bill.* 5. The Little Owl is much fmaller than the others; its plumage brown, regularly fpotted with white; its eyes pale yellow; its bill brown at the bafe, and yellow at the point; and may be called *The Little Brown Earlefs Owl, with Yellowifh Eyes, and a Brown and Orange Bill.*

Thefe characters apply in general; but, as in every other part of Nature, they are fometimes liable to confiderable variations, efpecially in the colours; enough, however, has been faid to diftinguifh them from each other. [A]

[A] The fpecific character of the Little Owl, *Strix Pafferina*, is "That its head is fmooth, and the feathers of its wings marked with five orders of fpots." It is very rare in England. In North America it is found from Hudfon's-bay to New York, and called by the Efquimaux *Shipmofpitl.* They build always in the pines, and in the middle of the tree, and lay two eggs; remain folitary in their retreat in the day, but are active in the fearch of their prey during the night.

FOREIGN BIRDS,

WHICH RESEMBLE THE OWLS.

I.

THE bird named *Caboor* by the Indians of Brazil, which has tufts of feathers on its head, and which is not larger than the Juniper Thrush. These two characters sufficiently shew it as a species of the Scops, if not a variety of the same species. Marcgrave is the only person who has described it, but he gives no figure of it; it is a kind of Owl, says he, of the size of a fieldfare; its head round; its bill short, yellow, and hooked, with two holes for the nostrils; the eyes beautiful, large, round, and yellow, with a black pupil; under the eyes, and on the side of the bill, are long brown hairs; the legs are short, and they, as well as the feet, are clothed completely with yellow feathers; the toes commonly four in number, with nails that are semilunar, black and sharp; the tail broad, the wings terminating at its origin; the body, the back, the wings, and the tail, are of a pale dusky colour, marked on the head and neck with very small white spots, and on the wings with larger spots of the same colour; the tail is waved with white; the breast and belly of a whitish-grey, clouded with light brown. Marcgrave adds,

that

that this bird is eafily tamed; that it can bend its head, and ftretch its neck fo much as to touch with the point of its bill the middle of its back; that it frolics with men like a monkey, and makes feveral antic motions; that it can erect the tufts on the fides of its head fo as to repre-fent fmall horns or ears; and that it feeds upon raw flefh. This defcription proves that it ap-proaches nearly to the European Scops; and I am almoft inclined to believe that the fame fpe-cies inhabits the Cape of Good Hope. Kolben informs us, that the Owls of the Cape are of the fame fize with thofe in Europe; that their fea-thers are partly red, partly black, with a mixture of grey fpots, which give them a beautiful ap-pearance; that feveral Europeans who live at the Cape tame them, and allow them to run about their houfes, and employ them for de-ftroying the mice. Though this defcription be not fo complete as that of Marcgrave, and does not warrant an abfolute conclufion, there is, however, a ftrong prefumption from the refem-blance of the properties of thefe birds, and from the fimilarity of the climates of Brazil and the Cape of Good Hope, that the two Owls are of the fame fpecies. [A]

[A] The *Cabure* is the *Brafilian-eared Owl* of Latham; the *Strix Brafilienfis* of Gmelin; the *Afio Brafilienfis* of Briffon; the *Noctua Brafilienfis* of Ray; and the *Ulula Brafilienfis* of Klein. Its fpecific character: " Its head eared; its body dufky-ferruginous, fpotted with white; below whitifh, with dufky ferruginous fpots; the tail-feathers ftriped with white."

II. The

II.

The bird of Hudson's Bay, called in that part of America *Caparacoch*; of which Edwards has given an excellent description and figure, and which he has named, *The Little Hawk-Owl*, because it participates of the nature of both these birds, and seems to be an intermediate shade. It is scarcely larger than the *Sparrow-hawk*, and the length of its wings and tail give it a similar appearance. The shape of its head and feet however shews, that it is more nearly allied to the genus of Owls; but it flies and catches its prey in broad day, like the other rapacious diurnal birds. Its bill is like that of the Sparrow-hawk, but not cornered on the sides; it is glossy and orange-coloured, covered almost entirely with hairs, or rather small ragged grey feathers, like most of the Owls; the iris is orange, the eyes encircled with white, and shaded with a little brown, speckled with small longish dusky spots, and on the outside of this white space is a black ring, which extends as far as the ears; beyond this black circle there is again some white; the crown of the head is deep brown, mottled with small white round spots; the arch of the neck and its feathers, as far as the middle of the back, are of a dull brown, edged with white; the wings are brown, and elegantly spotted with white; the scapular feathers are barred transversely with white and brown; the three feathers next the

the body are not spotted, but only bordered with white; the lower part of the back and the rump are of a deep brown, with transverse stripes of lighter brown; the lower part of the throat, the breast, the belly, the sides, the legs, the rump, and inferior coverts of the tail, and the smaller inferior coverts of the wings, are white, with brown transverse ribs, but the larger coverts of the wings are of an obscure ash-colour, with white spots on the two edges; the first of the quill-feathers of the wing is entirely brown without the least spot or border of white, and is not in the least like the rest of the quill-feathers, as may be remarked also in the other owls; the feathers of the tail are twelve in number, of an ash-colour below, and a dull brown above, with white narrow transverse bars; the legs and feet are covered with fine soft feathers, white like those of the belly, barred with shorter and narrower brown lines; the nails are hooked, sharp, and of a deep brown colour. [A]

Another individual of the same kind was a little larger, and its colours more dilute, which affords a presumption, that what has been described is a male, and the other a female. They were brought from Hudson's Bay to Edwards, by Light.

[A] This is *Strix Funerea* of Linnæus and Muller, and the *Strix Canadensis* of Brisson, and the *Canada Owl* of Latham. It flies high like a hawk, and preys by day upon the White Grous. It attends the fowler, and often steals the game before he has time to pick it up. It is found in North America, in Denmark, and Sweden, and is very frequent in Siberia.

III.

The HARFANG.

This bird inhabits the northern parts of both continents, and is known by this name in Sweden. It is not furnifhed with tufts on the head, and it is ftill larger than the Great-eared Owl. Like moft northern birds, its colour is fnowy-white. But we fhall borrow the excellent defcription which Edwards has given of this rare bird, which we could not procure.

" The Great White Owl," fays this author, " is one of the largeft of the Nocturnal Birds " of Prey, and at the fame time it is the moft " beautiful, for its plumage is white as fnow: " its head is not fo large in proportion as that " of the Owls; its wings when fpread, meafure " fixteen inches from the fhoulder to the end of " the longeft feather, which may give an idea of " its bulk. It is faid to prey in open day upon " the White Grous about Hudfon's Bay, where " it remains the whole year. Its bill is hooked " like a hawk's, and has no corners on the edges; " it is black, and perforated with wide noftrils, and " is alfo almoft entirely covered with ftiff feathers, " fimilar to the briftles at the bafe of the bill, and " reflected outwards. The pupil is encircled by " a brilliant-yellow iris; the head, the body, " the wings, and the tail are marked with fmall

" brown

" brown spots. The higher part of the back
" is transversely barred with some brown lines,
" the sides below the wings are also barred in
" the same manner, but by narrower and lighter
" lines : the great feathers of the wings are spot-
" ted with brown on their outer edges; there
" are spots also on the coverts of the wings,
" but the inferior coverts are pure white. The
" legs and feet are covered with white feathers ;
" the nails are long, strong, black, and very
" sharp." " I have another specimen of the same
" bird, (Edwards subjoins,) " in which the spots
" are more frequent, and the colour more in-
" tense."

This bird is common in the country about
Hudson's Bay ; but it seems to be confined to
the northern tracts ; for in the New Continent
it is very rare; in Pennsylvania and in Europe it
never appears farther south than Dantzick. It
is almost white, and spotless in the mountains of
Lapland. Klein informs us, that it is named
Hûrfang in Sweden, and *Weissebunte Schlictete-
eule* (i. e. White-chequered Smooth Owl) in Germany, and that he had in Dantzick a male and
female alive for several months in 1747*. Ellis

* *White Owl with earthy spots*. *Hûrfang*, Swed.; *Weissebunte
Schlichtete-eule*, Germ. On the 3d January 1747, I gave a stuffed
specimen to the cabinet of the Society of Gûar. When alive it
weighed three pounds and a half. The length from the point of
the bill to the end of the tail was one ell and a sixteenth, the
alar extent two and three-fifths; the bill and nails black ; the
cheeks, the lower part of the wings, the rump and the legs covered
with a milky shag ; the upper part of the body marbled with white
and cinereous.

relates

relates that this bird and the Great-eared Owl are frequent in the tracts near Hudson's Bay : it is, says he, of a dazzling white, hardly distinguishable from snow; it appears the whole year, flies often in open day, and hunts white partridges (*grous*). On the whole, therefore, the Harfang, which is the largest of all the Owls, is most frequent in the northern regions *, and probably avoids the heats of the south. [A]

* We have seen that it inhabits Lapland, Sweden, and the North of Germany : it is also found in Pennsylvania and Hudson's Bay, probably in Iceland; for Anderson has given a figure of it in his description of Iceland; and though Horrebow, who has criticised that work, asserts that no kind of Owl is found there, yet this ought not to be admitted upon the single credit of one whose principal aim it seems is to contradict Anderson.

[A] This is the *Strix Nyctea* of Linnæus, &c. the *Ulula Alba* of Klein, the *White Owl of Hudson's Bay* of Brisson, the *Great White Owl* of Edwards, and the *Snowy Owl* of Pennant and Latham.— The specific character, "the head smooth, the body whitish, with dusky lunar spots dispersed."

IV.
The CAYENNE OWL†.

Le Chat-huant de Cayenne, Buff.
Strix Cayanensis, Gmel.

This bird has been described by no naturalist. It is of the size of the Tawny Owl, from

† Specific character : " The body striated with rufous, waved transversely with dusky colour; the irides yellow."

which

which it differs by the colour of its eyes, which
are yellow; so that it is perhaps equally related
to the White Owl, but really differs from both.
It is particularly remarkable for its rufous plumage, waved transversely with brown narrow
lines, not only on the breast and belly, but even
on the back; its bill is of a flesh colour, and
its nails black.—This description, with the in-
spection of the figure, will be sufficient to recog-
nise it.

V.

The CANADA OWL*.

La Chouette, ou *Grande Chevêche de Canada*, Buff.
Strix Funerea, Linn.

This is considered by Brisson as a specimen of the Tawny Owl, but it appears to be
more allied to the Brown. It differs from the
latter, however, because its breast and belly are
marked with regular brown cross bars; and this
singular property is also observed in the Little
Owl of America.

* Specific character:—" The head smooth, the body dusky
spotted with white, streaked below with white and dusky, the wing
feathers variegated with white spots, the tail feathers streaked with
white." Its length seventeen inches, and its alar extent two feet.
It weighs twelve ounces.

VI. The

VI.

The SAINT DOMINGO OWL*.

La Chouette, ou *Grande Chevêche de Saint-Domingue*, Buff.
Strix Dominicenfis, Gmel.

This bird was fent us from St. Domingo,
and feems entirely a new fpecies.   It is the
neareft related to the Brown European Owl.
Its bill is larger, ftronger, more hooked than
that of any other Earlefs Owl.   It differs from
the Brown Owl in another circumftance alfo;
its belly is of a rufty uniform colour, and there
are only fome longitudinal fpots on the breaft;
whereas the Brown Owl of Europe is marked
on the breaft and belly with large oblong pointed fpots, which has given occafion to the name
of *Flaming Owl*.   *Noctua flammeata.*

* Specific character :—" The head fmooth, the abdomen rufous,
the breaft marked with ftraggling longitudinal fpots."

BIRDS

WHICH HAVE NOT THE POWER OF FLYING.

FROM the light birds which foar in the region of the clouds, we pafs to thofe that are borne down by their weight, and cannot rife from the furface. Our tranfition is fudden; but knowledge is acquired in the mode of comparifon, and the oppofition and contraft will throw additional light on the hiftory of the winged race. Indeed, without examining clofely the end of the chain, we cannot diftinguifh the intermediate links. When Nature is difplayed in her whole extent, fhe prefents a boundlefs field, where the various orders of being are connected by a perpetual fucceffion of contiguous and refembling objects: but it is not a fimple uniform feries, it ramifies at intervals in all directions; the branches from different parts bend, and run into each other, and thefe flexions and this tendency to unite, are moft remarkable near the extremes. We have feen in the clafs of quadrupeds, that one end of the chain ftretches to the tribe of birds in the different kinds of bats, which like thefe have the power of flying. The other end of the chain, we have perceived, defcends to the order of whales, in the feal, the

10   *wallrus,*

wallrus, and the *manati*: another branch was obferved rifing from the middle, and connecting
the monkey to man by the intermediate links of
the baboon, the pigmy-ape, and the orang-utang.
On the one fide, a fhoot bending through the
ant-eaters, the *phatagins*, and the *pangolins*, which
refemble in fhape the crocodiles, the *inguana*, and
the lizzards, unites the reptiles to the quadrupeds ; on the other, through the *tatous*, whofe
body is completely fheathed in a bony covering,
it approaches the cruftaceous animals.   It will
be the fame with refpect to the band which
connects the numerous order of birds ; if we
place its origin in thofe birds which fhoot nim-
bly with light pinions through the mid-way air,
it will gradually pafs through various minute
fhades, and at laft terminate in thofe which are
oppreffed with their weight, and deftitute of the
inftruments neceffary to impel their aërial courfe.
The lower extremity will be found to divide
into two branches ; the one containing terreftrial
birds, as the Oftrich, the Touyou, the Caffowary, and the Dodo, which cannot rife from the
ground ; the other including the Pinguins and
other aquatic birds, which are denied the ufe,
or rather the refidence of earth and air, and
which never leave the furface of the water,
their proper element.   Such are the ends of
the chain ; and we ought to examine thefe
with attention before we venture to furvey
the intermediate links, in which the proper-

ties

ties of the extremes are variously blended.
To place this metaphysical view in its proper
light, and to elucidate the ideas by actual examples, we ought, immediately after treating
of quadrupeds, to begin the Natural History of
the Birds which are the nearest related to these
animals. The Ostrich resembling the camel in
the shape of its legs, and the porcupine in the
pipes or prickles with which its wings are armed, ought to be ranged next the quadrupeds.
But philosophy must often yield to popular opinions; the numerous herd of naturalists would
exclaim against this classification, and would regard it as an absurd innovation, proceeding
merely from the love of singularity and contra-
tradiction. But besides the general resemblance
in size and outward appearance, which alone
ought to place it at the head of the winged race,
we shall find that there are many other analogies
to be found in the internal structure ; and that
being almost equally related to the birds and to
the quadrupeds, it must be considered as the
intermediate shade.

In each series or chain which connects the
universal system of animated nature, the branches
which extend to the subordinate classes are always short, and form very small *genera*. The
birds that are not fitted to fly, consist only of
seven or eight species ; the quadrupeds that are
able to fly, amount but to five or six. The
same remark may be applied to the other lateral

ramifications. These are fugitive traces of nature, which mark the extent of her power, which set defiance to the shackles of our systems, and burst from the confinement of our narrow circle of ideas.

THE OSTRICH

The OSTRICH*.

L'Autruche, Buff.
Struthio Camelus, Linn. Gmel. Will. Briff. &c.
The Black Oftrich, Alb. Sparr. Lath. &c.

THE Oftrich was known in the remoteft ages,
and mentioned in the moft ancient books.
It is frequently the fubject from which the facred writers draw their comparifons and allego-
ries †. In ftill more diftant periods, its flefh

* The Greek appellation Στρεθοκαμηλος, or fimply Στιεθος, is derived from ςρεθος, which fignifies a *fparrow,* or a bird in general ;
and καμηλος, a *camel* ; on account of the refemblance which the
Oftrich bears to that quadruped. The fame terms were introduced into Latin ; *Struthocamelus,* and fometimes *Struthio.* In Hebrew it was called *Jacuah* ; in Arabic, *Neamah* ; in Spanifh it is
now termed *Ave-Struz* ; in Italian, *Strutzo* ; in German, *Straufs.*

Linnæus ranges the Oftrich, the Galeated Caffowary, and the
Touyou, in the fame genus *Struthio* among the *Gallinæ.* The fpecific character of the Oftrich is, that it has three toes. Mr. Latham
in his laft work, *Index Ornithologicus,* has very properly formed another order, that of *Struthiones,* inferted after the *Grallæ,* and which
contains the *genera* of the Dodo, the Toyou, the Caffowary, and
the Oftrich. The character of the laft : that its *bill* is ftraight,
depreffed and rounded at the end ; the *wings* fhort, and ufelefs for
flying ; the *thighs* naked above the knees ; and two toes both turned
forward.

† *Oftriches fhall dwell there, and the Satyrs fhall dance there.*
ISAIAH, xiii. 21.

*Even dragons draw out the breafts, and give fuck to their young ;
but the daughters of my people become cruel like the Oftriches in the
wildernefs.* LAMENTAT. iv. 3.

*I will make lamentation like the dragons, and mourning like the
Oftriches.* MICAH, i. 8.

feems

feems to have been commonly ufed for food;
for the legiflature of the Jews prohibits it as un-
clean *. It occurs alfo in Herodotus †, the moft
antient of profane hiftorians, and in the writings
of the firft philofophers who have treated of the
hiftory of Nature: how indeed could an animal
fo remarkably large, fo ftrangely fhaped, and fo
wonderfully prolific, and peculiarly fitted for
the climate, as the Oftrich, remain unknown
in Africa and part of Afia, countries peopled
from the earlieft ages, full of deferts indeed, but
where there is not a fpot that has not been
trodden by the foot of man?

The family of the Oftrich, therefore, is of
great antiquity; nor in the courfe of ages has

---

* " And thefe alfo fhall have abomination among fowls. . . .
the Oftrich alfo, and the Cormorant," &c. Levit. xi. 13 & 16.
" But thefe are they whereof you fhall not eat. . . . nor the Oftrich, nor the night-crow," &c. Deut. xiv. 12 & 15.

† Salerne is of opinion that Herodotus fpeaks of three kinds of
Στρθοι: the *aquatic*, or *marine*, which is the fifh called plaice; the
aerial, which is the fparrow; and *the terreftrial* (Καλαγαιος) which is
the Oftrich. I can difcover only the laft, and I fhould render the
epithet Καλαγαιος differently, *fubterranean*; not that I believe in the
exiftence of fuch Oftriches; but Herodotus is there defcribing the
fingular productions peculiar to a certain region of Africa. The
common Oftrich was unlikely to be felected, fince the Greeks knew
it was common in Africa; and the fancy or credulity of the ancient hiftorian might create or affume thofe ideal beings.
Nor is it probable that fo rich, fo precife, and fo finifhed a language as the Greek, would affign the generic name of Oftrich to a
bird or a fifh. If I were allowed to offer a conjecture, I fhould fay
that the *Aerial Struthos* was the Long-tailed Oftrich, which in feveral parts of Africa is at prefent called *The Flying Oftrich*; and I
fhould fuppofe that the *Aquatic Struthos* was fome heavy water-fowl
whofe wings were unfit for flying.

it

it varied or degenerated from its native purity.
It has always remained on its paternal eftate;
and its luftre has been tranfmitted unfullied by
foreign intercourfe. In fhort, it is among the
birds what the elephant is among the quadrupeds, a diftinct race, widely feparated from all
the others by characters as ftriking as they are
invariable.

The Oftrich is reckoned the largeft of the
birds; but it is deprived of the prerogative of
the winged tribe, the power of flying. The one
which Vallifnieri examined weighed, though it
was very lean, fifty-five pounds, after the entrails were taken out; fo that, allowing twenty
pounds for thefe, and the fat that was wanting *,
we may eftimate the weight of an Oftrich when
alive, and in tolerable habit, at feventy-five or
eighty pounds. With what amazing force, then,
muft the wings, and the impelling mufcles of thefe
wings, have been endowed, to have been able to
raife and fufpend in the air fo huge a mafs? The
power of Nature appears to the fuperficial obferver as infinite; but when we examine clofely
the minute parts, we perceive that every thing is
limited; and to difcriminate with accuracy thefe
limits, which the wifdom, and not the weaknefs,
of Nature has prefcribed, is the beft method to

---

* Its two ftomachs, after being properly cleaned, weighed only
fix pounds; the heart, with the auricles, and the trunks of the large
veffels, was one pound feven ounces; the two pancreafes one pound;
the inteftines, which are very long and thick, muft be of confiderable weight.

ftudy

study her works and operations. In the present
case, the weight of seventy-five pounds exceeds
all the exertions of animal force to support it in
the medium of the atmosphere. Other birds
also which approach in size to the Ostrich, such
as the *Thuiou*, the Cassowary, and the Dodo, are
held down to the surface of the earth : but their
weight is not the sole obstacle ; the strength of
the pectoral muscles, the expansion of the wings,
their favourable insertion, the stiffness of the
quill-feathers, &c. would here be conditions the
more necessary, as the resistance to be overcome
is greater : but these requisites are entirely wanting ; for, to confine myself to the Ostrich, this
bird has, properly speaking, no wings ; since
the feathers inserted in the shoulders, instead of
forming a compact body fit to make a powerful
impression upon the air, are divided into loose
silky filaments, and the feathers of the tail are
of the same downy texture ; nor can they admit the varying positions which are necessary
for regulating their course. It is remarkable
that in the Ostrich the feathers are all of the
same texture ; whereas in most other birds, the
plumage is composed of different kinds of feathers. Those next the skin are soft and woolly ;
the coverts are closer and more solid ; and the
quill-feathers, which are destined to perform the
motions, are long and stiff. The Ostrich is,
therefore, confined to the ground by a double
chain ; by its great weight, and the structure of

its

its wings, it is condemned, like the quadrupeds,
to traverfe with labour the furface, and exiled
from the region of the air; and in both external and internal ftructure it bears great refemblance to thefe animals; like them, the greateft
part of its body is covered with hair rather than
feathers; its head and fides are almoft naked;
and its legs, in which its ftrength chiefly confifts, are thick and mufcular; its feet are ftrong
and flefhy, refembling thofe of the camel, which
differs from the other quadrupeds in that refpect; its wings, furnifhed with two pikes like
thofe of the porcupine, are to be regarded rather as
a kind of arms deftined for its defence; the orifice of the ear is uncovered, and only lined with
hair in the infide at the auditory canal; its upper eye-lid is moveable, as in almoft all the quadrupeds, and is edged with long eye-lafhes as in
man, and in the elephant; the general ftructure
of the eyes is moft analogous to what obtains in
man, and they are fo placed that both of them
point to the fame object. The parts near the
bottom of the *fternum*, and near the *os pubis*,
which, as in the camel, are callous, and deftitute
of hair or feathers, indicate its weight, and reduce it to a level with the humbleft of the beafts
of burden. Thevenot was fo ftruck with the
analogy between the Oftrich and the Drome-
dary, that he fancied he could perceive the
hump on its back; but, though the back is indeed arched, there is nothing fimilar to the flefhy

Y 4 protu-

protuberance that occurs in camels and drome-
daries.

If we proceed, from the furvey of its external
form, to examine its internal ſtruɑure, we ſhall
difcover other properties which diſtinguiſh it
from the birds, and new analogies which link it
with the quadrupeds.

The head is very fmall *, flat, and compofed
of foft tender bones †, but the crown is hard-
ened by a plate of horn. It is fupported in a
horizontal fituation by a bony column near three
feet in height, confiſting of feventeen *vertebræ*.
The body is commonly kept in the direction parallel to the horizon; the back is two
feet long, formed by feven *vertebræ*, and with
thefe are articulated on each fide feven ribs, two
falfe and five true; the laſt being double at
their origin, and afterwards uniting into a fingle
branch. A third pair of falfe ribs form the *cla-
vicle*; and the five true ribs are conneɑed by
cartilaginous ligaments to the *ſternum*, which
defcends not to the lower belly as in moſt birds,
and which is lefs projeɑing; it refembles a
buckler in ſhape, and is broader than even the
ſternum of a man. From the *os facrum* arifes a

* Scaliger remarks that many ponderous birds, fuch as the common cock, the peacock, the turkey, &c. have alfo a fmall head; but
that moſt birds which excel in flight, whatever be their fize, are furniſhed with a proportionally bigger head. Scal. *Exercit. in Car-
danum.*

† The anatomiſts of the Academy found a fraɑure in the cranium
of one of the fubjeɑs which they diffeɑed.

kind

kind of tail, confifting of feven *vertebræ*, fimilar
to thofe in man; the *os femoris* is a foot long;
the *tibia*, and *tarfus*, a foot and a half each; every
toe confifts of three *phalanges* as in man, while
other birds have feldom an equal number *.

The bill is rather fmall †, but opens wide;
the tongue is very fhort, and deftitute of *papillæ*.
The *pharynx* is broad, proportioned to the aperture of the mouth, and would admit a body of
the fize of the fift. The *æfophagus* is alfo wide
and ftrong, and terminates in the firft ventricle,
which in this bird performs three different functions; that of a craw, becaufe it is the firft; that
of a ventricle, being partly mufcular, and partly
confifting of longitudinal circular fibres; and
that of the glandulous protuberance, which generally occurs in the lower part of the *æfophagus* next the gizzard, fince it is furnifhed with
a great number of glands, conglomerated, and
not conglobated, as in moft other birds ‡. The
firft ventricle is fituated below the fecond; fo
that what is generally termed the *fuperior orifice*, in regard to its place, is in this cafe really
the inferior. The fecond ventricle is often divided from the firft by a flight conftricture; and
fometimes it is befides formed into two cavities

---

* Paré and Vallifnieri.

† Briffon fays, that the bill is unguiculated; Vallifnieri, that it
is pointed obtufely without any hook. The tongue varies much in
different fubjects.

‡ *Mem. pour fervir à l'Hiftoire des Animaux.*

by

by a fimilar conftricture ; but this divifion can never be perceived externally. It is covered with glands, and invefted with a villous coat fomething like flannel, but with little adhefion, and perforated with an infinite number of fmall holes, correfponding to the orifices of the glands. It is not fo ftrong as the gizzards of birds generally are ; but it is ftrengthened externally by very powerful mufcles, fome of them three inches thick. Its outward form refembles much that of the human ventricle.

Du Verney pretends that the hepatic duct terminates in this fecond ventricle *, as happens in the tench, and many other fifhes, and fometimes even in man, according to the obfervation of Galen †. But Ranby ‡ and Vallifnieri affirm, that in feveral Oftriches which they examined, they always found the infertion of this duct in the *duodenum* two inches, one inch, and fometimes only half an inch below the *pylorus*. Vallifnieri alfo points out the origin of this miftake, if it be fuch, adding that in two Oftriches he traced a veffel from the fecond ventricle to the liver, which he firft took for a branch of the hepatic duct, but afterwards difcovered that it was an artery which conveyed blood to the liver, and not bile to the ventricle.

* *Hift. de l'Academie des Sciences,* 1694.
† In Vallifnieri.
‡ Philofoph. Tranfact. N° 386.

The

The *pylorus* varies in regard to its width in different subjects; it is generally tinged with yellow, and, as well as the cavity of the second ventricle, is imbued with a bitter liquor. This is easily accounted for, because the hepatic duct takes its origin in the *duodenum*, and runs upwards.

The *pylorus* discharges itself into the *duodenum*, the narrowest of all the intestines, and in which are also inserted the two pancreatic ducts, a foot, and sometimes two or three feet, below the junction of the hepatic; while in other birds the insertion is made close to the gall duct.

The *duodenum* and the *jejunum* are without valves; the *ileon* is furnished with some, as it runs into the *colon*. These three small intestines are nearly half the length of the whole alimentary canal, which, in different subjects even of the same bulk, is subject to variation, being sixty feet in some, and only twenty-nine in others.

The two *cæca* rise from the beginning of the *colon*, according to the anatomists of the Academy; or from the end of the *ileum*, according to Ranby. Each *cæcum* forms a kind of hollow cone two or three feet long, an inch wide at the base, and furnished in the inside with a valve in the form of a spiral plate, making near twenty revolutions from the bottom to the top, as in the hare, the rabbit, the sea-fox, the ray, the cramp-fish, and the thornback, &c.

The

The *colon* alfo is furnifhed with leaf-fhaped valves, but which, inftead of turning fpirally, form a crefcent that occupies rather more than half the circumference of the *colon :* fo that the ends of the oppofite crefcents flightly overlap each other. And this ftructure alfo occurs in the *colon* of the monkey, and in the *jejunum* of man, and marks the inteftine exteriorly with tranverfe parallel furrows, about half an inch diftant, and correfponding to the interior valves : but it is remarkable that thefe crefcents do not occur through the whole length of the *colon,* or rather that the Oftrich has two very different *colons ;* the one broad and about a foot long, furnifhed with leafy valves ; the other, narrower, and totally deftitute of valves, but extending to the *rectum.*

The *rectum* is very wide, about a foot long, and near its termination covered with flefhy fibres. It opens into a large bag or bladder confifting of membranes, the fame as the inteftines, but thicker, and fometimes containing even eight ounces of urine*. For the ureters make their difcharge by a very oblique infertion, as in the bladder of land animals; and not only convey

---

* The urine of the Oftrich difcharges ink-fpots according to Hirmolaüs. The affertion may be falfe, but Gefner was miftaken in contradicting it, on the ground that no bird has urine. If this were admitted, of what ufe are the kidneys, and the ureters ? The only difference in this refpect between birds and quadrupeds is, that in the former the bladder opens into the *rectum.*

urine,

urine, but alfo the white glutinous matter that accompanies or envelopes the excrement in all birds.

This firft bag, which wants only the neck to be a real bladder, communicates by an orifice furnifhed with a kind of *fphincter* with the fecond and laft bag, which is fmaller, and ferves for the paffage of the urine and the folid excrements; it is almoft fhut by a cartilaginous nut, adhering at its bafe to the junction of the *os pubis*, and cleft in the middle like that of the apricot.

The folid excrements are very like thofe of fheep and goats, being divided into little balls, whofe bulk bears no relation to the capacity of the inteftines where they are formed. In the fmall inteftines, the appearance is like that of foup, fometimes green, fometimes black, according to the quantity of aliment, which acquires confiftence as it approaches the thick inteftines, but does not receive its fhape until it enters the fecond *colon*.

Near the *anus* are fometimes found fmall facs, fomewhat fimilar to what occur in the fame parts in lions and tigers.

The mefentery is tranfparent through its whole extent, and in fome parts it is a foot broad. Vallifnieri pretends to have difcovered in it manifeft traces of lymphatic veffels: Ranby alfo fays, that the veffels of the mefentery are very diftinct, but adds, that its glands can hardly be perceived.

To

To moſt obſervers indeed they have been altogether inviſible.

The liver is divided into two great lobes, as
in man, but it is placed nearer the middle of
the hypochondriac region, and has no gall bladder. The ſpleen is contiguous to the firſt ſtomach, and weighs at leaſt two ounces.

The kidneys are very large, ſeldom parted into
ſeveral lobes as in other birds, but oftener ſhaped
like a guitar, with a broad belly.

The ureters never creep along the kidneys
as in moſt other birds, but penetrate into their
ſubſtance.

The *epiploon* is very ſmall, and only covers a
part of the ventricle ; but in its ſtead we find over
all the belly, and ſometimes on the inteſtines, a
coat of fat or tallow ſpread between the *aponeu-
roſes* of the muſcles of the abdomen, and from
two to ſix inches thick. It was this fat mixed
with blood that formed the *manteca*, which was
highly eſteemed, and extremely dear among the
Romans, who, according to Pliny, reckoned it
more efficacious than gooſe fat, for rheumatiſm,
cold ſwellings, and palſy ; and even at preſent
the Arabians preſcribe it in theſe diſorders*. Valliſnieri is the only one who, probably from his
happening to diſſect very lean Oſtriches, ſuſpects
the exiſtence of this fact ; and the more ſo, that in
Italy the leanneſs of the Oſtrich has paſſed into

* *The World Diſplayed*, vol. xiii.

a pro-

a proverb, *magro comme uno ftruzzo* * ; he adds,
that thofe which he examined appeared after
diffection like mere fkeletons: but this muft be
the cafe with refpect to all Oftriches that have
no fat, or in which it has been feparated, fince
there is no flefh on the breaft or belly, for the
mufcles of the abdomen do not become flefhy
till they reach the fides.

If from the organs of digeftion we pafs to
thofe of generation, we fhall find other analogies
to the ftructure which obtains in quadrupeds.
In the greateft number of birds, the *penis* is concealed; but in the Oftrich it is apparent, and of
a confiderable fize, compofed of two white ligaments, that are folid and nervous, four lines
diameter, and fheathed in a thick membrane,
and which only unite at the breadth of two
fingers from their extremity. Sometimes we
alfo meet in the fame part a red fpongy fubftance, fupplied with a multitude of veffels, and
very fimilar to the *corpus cavernofum* that is obferved in the land animals. The whole is inclofed in a common membrane, whofe texture is
the fame as that of the ligaments, though not
fo thick or hard. This *penis* is furnifhed with
neither gland nor prepuce; nor, according to
the anatomifts of the academy, is even perforated for the ejection of the feminal fluid;
but Warren pretends that he diffected an
Oftrich, whofe yard was five inches and a half

* " Meagre as an Oftrich."

long,

long, and furrowed longitudinally along the upper surface, with a kind of channel, which appeared to him to be deftined for conveying the
femen. Whether this channel was formed by
the junction of the two ligaments ; or that Warren miftook for the *penis* the cartilaginous nut
of the fecond bag of the *rectum*, which is in fact
parted ; or that the ftructure and fhape of this
organ is liable to vary in different fubjects ; it
appears that the yard adheres at its origin to the
cartilaginous nut, and bending downwards, it
paffes through the fmall fac, and emerges at the
external orifice, which is the *anus*, and which
being edged with a membranous fold, forms
at this part a falfe prepuce, that Dr. Brown has
undoubtedly miftaken for a real prepuce, for he
is the only perfon who afferts that the Oftrich
has that excrefcence *.

There are four mufcles attached to the *anus*
and the yard, whence refults a fympathy of motion ; and this is the reafon why, when the
animal voids its excrements, the yard protrudes
feveral inches †.

The tefticles differ widely in regard to fize in
different individuals, and vary even in the proportion of forty-eight to one ; owing, doubtlefs, to
their age, the feafon, the nature of the difeafe which
preceded death, &c. Their external fhape varies

---

* *Collect. Philof.*

† Warren learned this fact from perfons who kept feveral
Oftriches in England.

alfo,

alfo, but their external ftructure is always uniform; they lie on the kidneys, nearer the left
than the right. Warren imagined he could perceive feminal veficules.

The females alfo have tefticles; for fo we ought
to call thofe glandulous bodies, four lines in diameter and eighteen long, which are found under the *ovarium* adhering to the *aörta* and *vena
cava*, and which nothing but the predilection
of fyftem could convert into the lower glands of
the kidney. The female Little Buftard is alfo
furnifhed with tefticles fimilar to thofe of the
male, and there is reafon to believe that the female of the Great Buftard has the fame ftructure;
and if the Anatomifts of the Academy, in their
numerous diffections, have fuppofed that they
never met with any but males, it is becaufe they
would not admit an animal in which they found
tefticles to be a female. But every body knows
that the Buftard approaches the neareft of the
European birds to the Oftrich, and therefore all
that I have faid on the fubject of the generation
of tefticles in the bodies of female quadrupeds,
applies readily to this clafs, and will afterwards
perhaps be difcovered to admit of a greater extenfion.

Below thefe two glandulous bodies is placed
the *ovarium*, adhering alfo to the great blood
veffels; it generally contains eggs of different
fizes, inclofed in their capfule like fmall glands,
and attached to the ovarium by their ftalks.

This *ovarium* is fingle, as in almoft all birds; and we may remark by the way that this affords another prefumption againft the opinion of thofe who maintain that the two glandulous bodies which occur in all the females of quadrupeds, reprefent the *ovarium*, which is a fingle organ; inftead of admitting that they are really tefticles, which muft be reckoned among the double parts, both in the males of birds and in thofe of qua-drupeds *.

The funnel of the *oviductus* opens below the *ovarium*, and fends off to the right and left two winged membranous appendices, which refemble thofe that occur at the end of the tube in land animals. The eggs which are feparated from the *ovarium* are received into this funnel, and conveyed along the *oviductus* to the laft inteftinal fac, where they are difcharged through an ori-fice, which, in its natural ftate, is only four lines in diameter, but its wrinkled furface expands and forms a dilatation proportional to the bulk of the eggs. All the inner coat of the *oviductus* is alfo full of wrinkles, or rather folds, as in the third and fourth ftomachs of the ruminating animals.

---

* The Flamingo is the only bird in which the Anatomifts of the Academy found two *ovaria*; and thefe are, according to them, no-thing more than two hard folid glandulous bodies, of which the left one is divided into feveral unequal globules. But this is a fingular inftance from which no general conclufion can be drawn.

Laftly,

Laftly, the fecond inteftinal bag has its cartilaginous nut in the female as well as in the
male; and this nut, which fometimes projects
more than half an inch from the *anus*, has a
fmall appendix three lines in length, thin and
incurvated, which the Anatomifts of the Academy take for a *clitoris*, and with the greater
probability, as the fame two mufcles that are inferted in the bafe of the yard in the males, are
alfo connected to the origin of this appendix in
the females.

I fhall not dwell on the defcription of the
organs of refpiration, fince they refemble almoft
entirely thofe of the other birds; confifting of
two lungs of a fpungy fubftance with ten air
cells, five on each fide, of which the fourth is
here the fmalleft, as ufual in all the bulky fpecies
of birds: thefe cells receive the air from the
lungs, with which they have very diftinct communications; but they muft alfo have communications with other parts, though lefs apparent; for when Vallifnieri blew into the
trachea-arteria, he obferved an inflation along
the thighs and wings, which indicates a ftructure
fimilar to that of the Pelican, in which Mery
perceived, under the infertion of the wing, and
between the thigh and the belly, membranous
bags, which were filled with air during expi-
ration, or when air was injected forcibly into
the *trachea-arteria*, and which probably furnifh
it to the cellular texture.

Z 2

Dr.

Dr. Brown pofitively afferts, that the Oftrich has no *epiglottis:* Perrault fuppofes the fame thing, fince he beftows on a certain mufcle the office of fhutting the *glottis*, by contracting the *larynx*. Warren affirms that he perceived a glottis in the fubject which he diffected ; and Vallifnieri reconciles thefe oppofite opinions, by faying that there is really no perfect *epiglottis*, but that the pofterior part of the tongue fupplies the defect, clofing on the *glottis* in deglutition.

There are various opinions alfo with refpect to the number and form of the cartilaginous rings of the *larynx:* Vallifnieri reckons only two hundred and eighteen, and maintains, with Perrault, that they are all entire : Warren found two hundred and twenty-fix complete, exclufive of the firft ones which were imperfect, or thofe immediately under the forking of the *trachea.* All this may be true, confidering the great varieties to which the ftructure of the internal parts are fubject; but it proves at the fame time the rafhnefs of attempting to defcribe a whole fpecies from a fmall number of individuals, and the danger of miftaking anomalous varieties for conftant characters. Perrault obferved, that each of the two branches of the *trachea-arteria* is divided at its junction with the lungs into a number of membranous ramifications, as in the elephant. The brain, with its *cerebellum*, forms a mafs about two inches and a half long, and

8   twenty

twenty lines broad. Vallifnieri affirms, that the one he examined weighed only an ounce, which would not amount to the one-twelfth-hundredth part of the weight of the animal : he adds, that the ftructure was exactly fimilar to that of the brain of other birds, and precifely fuch as defcribed by Willis. I fhall obferve however, with the Anatomifts of the Academy, that the ten pairs of nerves arife and proceed from the *cranium*, in the fame manner as in land animals ; that the cortical and the medullary part of the *cerebellum* are alfo difpofed as in thefe animals ; and that we fometimes find the two vermiform apophyfes which occur in man, and a ventricle, fhaped like a writing pen, as in moft of the quadrupeds.

With refpect to the organs of circulation, I fhall only notice, that the heart is almoft round, while in other birds it is generally elongated.

In regard to the external fenfes, I have already defcribed the tongue, the ear, and the external form of the eye: I have only to add, that its internal ftructure is the fame with what is commonly obferved in birds. Ranby afferts, that the ball taken from its focket, fpontaneoufly affumes a form almoft triangular; he alfo remarks, that the quantity of the aqueous humour is greater, and that of the vitreous lefs than ordinary.

The noftrils are placed in the fuperior mandible, not far from its bafe, and on the middle of each aperture rifes a cartilaginous protuberance

covered

covered with a very thin membrane, and thefe apertures communicate with the palate by means of two canals which terminate in a pretty large cleft. We fhould be miftaken, were we to infer from the complicated ftructure of this organ, that the Oftrich poffeffed the fenfe of fmell in an eminent degree; for the moft un-doubted facts prove exactly the reverfe; and in general it appears that the chief impreffions, and the moft exquifite which this animal re-ceives, are thofe of fight, and of the fixth fenfe.

This fhort view of the internal organization of the Oftrich is more than fufficient to confirm the idea which I before gave, that this fingular animal muft be confidered as a being of an equivocal nature, and as forming the fhade be-tween the quadruped and the bird: and in a fyftem which would reprefent the true gradation of the univerfe, it fhould be referred neither to the clafs of quadrupeds nor to that of birds, but ranged in the intermediate place. Indeed, what rank can we affign to an animal whofe body is partly that of a bird, partly that of a quadruped*; its feet like thofe of a quadruped, its head fimi-lar to that of a bird; the male furnifhed with a *penis*, the female with a *clitoris*, as in the qua-drupeds; which is oviparous, and has a gizzard like the birds, and at the fame time is fupplied with feveral ftomachs, and with inteftines, whofe

* Ariftotle.

capacity

capacity and ſtructure are analogous partly to
the ruminating, and partly to the other quadrupeds?

In the order of fecundity, the Oſtrich ſeems
to be more nearly related to the quadrupeds than
to the birds; for its incubations are frequent,
and it hatches many young at a time. Ariſtotle
ſays, that, next after the Oſtrich, the bird which
he calls the *Atricapilla*, is that which lays the
moſt eggs; and he adds, that this bird, *Atricapilla*, lays twenty and more; whence it follows that the Oſtrich lays at leaſt twenty-five.
Beſides, the beſt informed modern hiſtorians and
travellers relate, that it has ſeveral ſittings, with
twelve or fifteen eggs in each. But if we refer
it to the claſs of the birds, it would be the largeſt,
and conſequently ought to be the leaſt prolific,
according to the law which Nature ſeems to have
conſtantly obſerved in the multiplication of animals, that it is univerſally proportional to the
bulk of the individuals; whereas, if we refer
it to the claſs of land animals, it appears diminutive beſide the largeſt ſpecies, and ſmaller than
thoſe of a middle ſize, as the hog, and its great
fecundity is therefore conſiſtent with the general
order of the univerſe.

Oppian, who entertained the ſtrange notion
that the camels of Bactriana copulated backwards, turning their tails to each other, believed
alſo that the *camel bird* (the name anciently applied to the Oſtrich) performs its embraces in

the

the fame manner; and he advances it as an un-
doubted fact. But this is no more probable with
refpect to the camel-bird than with refpect to
the camel itfelf; and though it is moft probable
that few obfervers have witneffed their coupling,
and that none have defcribed it, we ought ftill to
conclude, fince there is no proof of the contrary,
that it is accomplifhed in the ufual way.

The Oftriches are reckoned exceffively fala-
cious, and often copulate; and if we recollect
what has been already faid with refpect to the
dimenfions of the yard, we fhall readily conceive that this act is not performed by mere
compreffion, as in almoft all the other birds,
but that the male organ is really introduced into
the fexual parts of the female : Thevenot is the
only perfon who afferts that they pair, and that,
contrary to what is ufual with the large birds,
each male felects his female.

The time of laying their eggs depends on the
climate they inhabit, but is always near the fummer folftice; that is, about the beginning of July,
in the northern parts of Africa*, and towards
the end of December, in the fouthern tracts of
that continent †. The temperature of the climate has alfo great influence on the mode of
hatching. In the torrid zone, they are contented with depofiting their eggs in a heap of
fand loofely fcraped together with their feet, and

---

* Albert.   † Dampier.

leave

leave the developement of the young to the powerful agency of a burning fun. Nor is this always neceffary ; they are fometimes hatched, though neither covered by the mother, nor expofed to the influence of the folar rays *. But though the Oftrich has feldom or never recourfe to incubation, fhe is far from abandoning her eggs : fhe watches affiduoufly over their prefervation, and feldom lofes fight of them. This has given occafion to the faying, that they hatch them with their eyes ; and Diodorus relates a method of catching thefe animals, which is founded on their ftrong attachment to their expected offspring : this is to fet in the ground round the neft, at a proper height, ftakes, armed with fharp points, upon which the mother rufhes heedlefsly, and is transfixed.

Though the climate of France is much colder than that of Barbary, Oftriches have fometimes laid their eggs in the *Menagerie* of Verfailles ; but the Anatomifts of the Academy were unfuccefsful in their attempts to hatch them, either by artificial incubation, or by the heat of the fun, or by the application of the gradual heat of a flow regulated fire ; nor could they trace in any of the eggs the leaft marks of an incipient organization, or difcover the

---

* When Jannequin was at Senegal, he put two Oftrich's eggs into a cafk, and packed them well with tow ; but fometime afterwards, on opening it, he found that one of the eggs was quite ripe for exclufion.

flighteft indication of the production of a new being.—The yolk and the white of the one that had been heated by the fire, were only a little thickened; that expofed to the fun contracted a very putrid fmell; but neither fhewed the leaft appearance of the rudiments of a *fœtus;* and in fhort, this philofophical incubation was totally unfuccefsful*.—Reaumur had not yet appeared.

The eggs are extremely hard, heavy, and large; but fometimes they are fuppofed to be more bulky than they really are, thofe of the crocodile being miftaken for them †. It has been afferted, that they are as large as the head of a child ‡, that they would contain a quart ‖, that they weigh fifteen pounds, and that an Oftrich lays fifty § eggs in the year; Ælian goes as far as eighty. But moft of thefe circumftances are evidently exaggerated: for, firft, is it pof-fible that an egg, whofe fhell is not more than a pound in weight, and whofe capacity is at moft only a quart, could weigh fifteen pounds? To reconcile this, we muft fuppofe that the yolk and white are feven times denfer than water, three times than marble, and almoft as denfe as tin, which is rather a ftrained hypothefis. Secondly, Admitting with Willughby, that the Oftrich lays annually fifty eggs, weighing fifteen pounds each, it would follow that, in the courfe of the year, fhe would exclude feven hundred

* *Mem. pour fervir à l'Hift. des Anim.* † Belon.
‡ Willughby.   ‖ Belon.   § Leo Africanus.

and

and fifty pounds, which is too much for an animal that is fcarcely eighty pounds weight.

We muft therefore make a confiderable abatement both in the weight of the eggs, and in their number; but it is a pity that we have not fufficient data to afcertain the precife quantity. Ariftotle indeed renders it probable, that the number of eggs is from twenty-five to thirty; and the moft judicious modern writers ftate it at thirty-fix. If we fuppofe two or three layings in the year, and a dozen eggs to each, we might alfo admit the weight of each egg to be three or four pounds, allowing a pound for the fhell, and two or three for the white and yolk; but this is merely conjecture, and far from being accurate. Many people write, but few weigh, meafure, or compare. Of fifteen or fixteen Oftriches which have been diffected in different countries, only one has been weighed, and it is that which we have defcribed from Vallifnieri.—We are no better informed with refpect to the time neceffary for the incubation of the eggs; all we know, or rather all that is afferted is, that as foon as they are hatched, the young Oftriches are able to walk, and even to run and fearch for their food; inafmuch that in the torrid zone *, where they enjoy the proper degree of warmth, and can eafily provide their fuitable fubfiftence, they are emancipated at their birth, and abandoned by their mother, on

* Leo Africanus.

whofe

whofe affiftance they are independent. But, in the more temperate countries, as at the Cape of Good Hope *, the mother watches over her young fo long as her affiduous attention is neceffary; and in every climate her care is proportioned to their wants.

The young Oftriches are of an afh-gray the firft year, and entirely covered with feathers; but thefe are falfe, and foon drop. They are never reftored on the head, on the top of the neck, on the thighs, on the fides, and below the wings; but they are replaced on the reft of the body by plumes alternately black and white, and fometimes gray, from the blending thefe two colours into each other. The fhorteft are on the lower part of the neck; thofe on the belly and the back are longer; but the longeft of all are thofe at the extremity of the tail and of the wings, and are alfo the moft efteemed. Klein mentions, on the authority of Albert, that the dorfal feathers are very black in the males, and brown in the females; but the Academicians, who diffected eight Oftriches, five male and three female, found the plumage nearly alike in all; yet they never obferved red, green, blue, or yellow feathers, as Cardan feems to have believed, from a ftrange overfight in a work *De Subtilitate.*

Redi difcovered, from numerous obfervations,

---

* Kolben's Account of the Cape.

that

that almoſt all birds are ſubject to vermin in
their feathers, and even vermin of different
kinds ; that the greateſt number have inſects peculiar to them, and no where elſe found ; but
in no ſeaſon could he ever perceive them in
Oſtriches, though he examined a dozen of thoſe
animals, ſome of which had been recently brought
from Barbary. Further, Valliſnieri, who diſſected
two Oſtriches, found in the bowels neither *lum-
brici*, nor worms, nor inſects of any ſort. It
would ſeem therefore, that none of theſe creatures are fond of the Oſtrich fleſh ; that they
avoid it with an averſion, and that it has ſome
quality pernicious to their multiplication. Perhaps the breeding of vermin is prevented inter-
nally by the great powers of the ſtomach, and
the digeſtive organs. Many fabulous opinions
have been entertained on this ſubject. It has
been aſſerted, that the Oſtrich digeſts iron as
poultry digeſt grain ; and ſome authors have
even gone ſo far as to allege, that it could digeſt red-hot iron *. The laſt opinion requires
no ſerious refutation ; and it will be enough to
aſcertain from facts, if the Oſtrich can grind
down cold iron.

It is certain that theſe birds live chiefly on
vegetable ſubſtances ; that their gizzard is lined
with very ſtrong muſcles, as in all the granivo-

* Marmol, *Deſcrip. de l'Afrique.*

rous

rous clafs *; that they often fwallow bits of
iron †, copper, ftones, glafs, wood, and any
thing that occurs. I will not deny that they
may even fometimes fwallow hot iron, if the
quantity be fmall, and this perhaps without fuffering any inconvenience. It appears that they
fwallow whatever they can find, till their capacious ftomachs be completely filled ; and that the
need of ballafting them with a fufficient weight,
is one of the principal caufes of their voracity.
The gallinaceous tribe, and other granivorous
animals, whofe organs of tafte want fenfibility,
alfo fwallow many fmall ftones, when mixed
with their food, miftaking them probably for
grains ; but if ftones be offered alone, they will
perifh of hunger, and not touch one of them ;
and ftill lefs will they meddle with quick-lime.
We may therefore conclude, that the Oftrich is
one of the birds whofe fenfes of tafte and fmell
are the moft obtufe ; and in this circumftance
they are widely feparated from the quadrupeds.

But what become of thofe hard noxious fubftances, efpecially the copper, the glafs, and the
iron, which the Oftrich fwallows at random, and

* Though the Oftrich is actually omnivorous, it may ftill be
ranged in the granivorous clafs, fince in its deferts it lives on dates,
and other fruits, or vegetable fubftances ; and in *menageries* it may
be kept on the fame food. Alfo Strabo tells us, Book vi. that when
the hunters want to enfnare it, they ufe grain for bait.

† I fay often ; for Albert affirms that many Oftriches would not
fwallow iron, though they devoured hard bones, and even ftones
with avidity.

merely

merely with the view of repletion? On this fubject the authors are divided, and adduce particular facts in fupport of each opinion. Perrault,
having found feventy doubloons in the ftomach of
one of thefe animals, obferved, that moft of them
were worn down, and reduced to three-fourths
of their prominence. He conceived that this
was occafioned by their mutual friction, and the
comminution of pebbles, rather than by the action of any acid ; fince fome of thefe doubloons
were much corroded on the convex furface, which
was moft expofed to the attrition, and yet not
in the leaft affected on the concave fide. He
therefore concluded, that, in thefe birds, the folution of the food is not performed merely by
fubtile and penetrating juices, but is effected by
the organic action of the ftomach, which com-
preffes its aliments, and agitates them inceffantly
with thofe hard bodies which they inftinctively
fwallow. And, becaufe the contents of the ftomach were tinged with green, he inferred that
the copper was actually diffolved in it ; not by
any particular folvent, nor by the powers of digeftion, but in a fimilar manner to what would
take place if that metal were ground with herbage, or with fome acid or faline liquor. He
adds, that copper, far from affording nourifh-
ment in the ftomach of the Oftrich, really acts
as a poifon, and that all thofe who fwallowed·
much of it foon died.

Vallif-

Vallifnieri, on the other hand, imagines, that the Oftrich digefts or diffolves the hard fub-ftances chiefly by the action of the acid liquor of the ftomach ; but he does not exclude the effect of attrition which may affift the folution. His proofs are thefe :

1. Bits of wood, iron, or glafs, which have remained fome time in the ftomach of the Of-trich, are not fmooth and fhining, as they would be if worn by the friction of the coats ; but are rough, furrowed, perforated, and precifely fuch as would be produced by the corrofion of a fol-vent.

2. This folvent reduces the hardeft and the fofteft bodies alike to impalpable molecules, which may be obferved by the microfcope, and even by the naked eye.

3. He found in the ftomach of the Oftrich a nail fixed in one of the fides, in fuch a manner as to prevent its meeting the oppofite fide, and con-fequently the compreffion of the contents ; yet the food was as completely diffolved in this ven-tricle as in another in which the ufual action could take place ; and this at leaft proves, that in the Oftrich digeftion is not performed folely by tritu-ration.

4. He obferved a copper thimble in the fto-mach of a capon, which was corroded only where it touched the gizzard, and confequently where it was leaft expofed to the attrition of the hard

hard fubftances; whence the folution of metals in the gallinaceous tribe muft be afcribed rather to the action of fome menftruum, than to the preffure and refiftance of the coats; and the analogy naturally extends to the Oftrich.

5. He difcovered in the ftomach of the Oftrich, a piece of money which had been fo completely corroded, that its weight was reduced to three grains.

6. The glands of the firft ftomach exude, when fqueezed, a vifcous, yellowifh, infipid liquor, which, however, quickly marks iron with a dull fpot.

7. Laftly, the activity of thefe juices, the force of the mufcles of the gizzard, and the black colour which tinges the excrements of the Oftriches that have fwallowed iron, which alfo happens to thofe perfons who ufe martial preparations, and have a vigorous digeftion, fupporting the preceding facts, incline Vallifnieri to conjecture, not indeed that the Oftriches really digeft iron, and are nourifhed by it, as feveral infects and reptiles feed on earth and ftones, but that the ftones and the metals, efpecially iron, being diffolved by the gaftric liquor, temper, like abforbents, the acrid juices of the ftomach, and mix with the nutriment as ufeful ingredients for feafoning it, and increafing the action of the folids. And this conclufion is the more reafonable, fince iron is known to enter into the compofition of living beings; and,

when fufficiently attenuated by proper acids, it becomes volatile, and fhews a tendency as it were to vegetate, and affume forms analogous to thofe of plants, as is obferved in the *arbor Martis* [*].   In this fenfe only can the Oftrich be faid to digeft iron : and even admitting that the powers of the ftomach were fufficient to produce the effect, it is ftill extremely ridiculous to imagine, that the gizzard has the beneficial qualities of a medicine, and is proper for affifting a debilitated appetite.   But fuch is the nature of the human mind ! ftruck with a rare or fingular object, fhe never fails to heighten the wonder by the addition of chimerical, and often abfurd qualities.   Hence it has been affirmed, that the moft tranfparent ftones found in the ftomach of the Oftrich, have the virtue, when applied to the neck, of promoting digeftion ; that the inner coat of the gizzard has the power of correcting a cold temperament, and of rekindling the fire of luft ; its liver, that of curing the falling-ficknefs ; its blood, that of reftoring fight ; and, the fhell of its egg reduced to powder, that of relieving the gout, and the gravel, &c.   Vallifnieri had an opportunity of demonftrating by experiments the falfity of thefe pretended virtues ; and his experiments are the more decifive, as they were

---

[*] *Memoires de l'Academie des Sciences, années* 1705, 1706, *& fuivantes* ; VALLISNIERI. His opinion is farther confirmed by the obfervations of Santorini on bits of money, and keys found in the ftomach of an Oftrich which was diffected at Venice ; and by the experiments of the Academy *de Cimento* on the digeftion of birds.

per-

performed on perfons the moft credulous, and the moft fubject to prejudices.

The Oftrich is a bird peculiar to Africa, the iflands contiguous to that continent*, and that part of Afia which borders on it. Thefe regions, which are the native abodes of the camel, the rhinoceros, the elephant, and many other large animals, muft alfo be the paternal feat of the Oftrich, which is the elephant of the birds. They are very numerous, according to Dr. Pocock, on the mountains fituated on the fouthweft of Alexandria. A miffionary mentions that they occur at Goa, though much more rarely than in Arabia †. Philoftratus pretends that Apollonius found them even beyond the Ganges ‡ ; but this was certainly at a period when Hindoftan was lefs populous than at prefent. Modern travellers have never obferved any in that country, unlefs fuch as were carried thither § ; and all agree, that on either fide of the line, they never pafs beyond the thirty-fifth degree of latitude ; and as they cannot fly, they are in the fame predicament with the quadrupeds in the fouthern tracts of the ancient conti-

---

* The *Vorou-Patra* of Madagafcar is a kind of Oftrich which retires into folitary fpots, and lays eggs of an uncommon fize.

† Travels of Father Philip, a bare-footed Carmelite.

‡ Life of Apollonius, book iii.

§ They are kept in the *menageries* of the king of Perfia, according to Thevenot ; which fhews that they are not common in that country. On the road from Ifpahan to Schiras four Oftriches were brought into the *caravanfera*, fays Yemelli Carreri.

nent ;

nent; that is, they cannot migrate into the new continent. Hence they have never been dif-covered through the whole range of America, though the name has been applied to the Tou-you, which is analogous to them indeed, but is quite a different species, as we shall soon find. For the same reason they have never been found in Europe, though they might enjoy a climate suited to their nature in the Morea, in the south of Spain, and in Italy. But, before they could migrate into these countries, they must have crossed the intervening seas, which was impos-sible; or follow the line of coast as high as the fiftieth degree of latitude, another obstacle equally insurmountable for an animal that de-lights in the parched plains. The Ostriches pre-fer the most solitary and the most arid tracts, that are scarcely ever refreshed by genial showers *; and this confirms the account of the Arabians, that they do not drink. They assemble in these deserts in numerous flocks, which at a distance

---

* " Theophrastus says, that the Ostrich breeds in that part of Africa where it does not rain." GESNER. All the travellers and naturalists are agreed on this point. G. Warren is the only per-son who reckons the Ostrich aquatic: he admits that it never swims; but says that its legs are tall, and its neck long; so that it could wade in the water and catch its prey; and, as its head resembles that of a goose, hence he infers that it is a river bird. *Philos. Transf.* N° 394. Another hearing that in Abyssinia the Ostriches were of the size of an ass, and that their neck and feet resembled those of the quadrupeds, concluded that they must have an ass's tail and feet. SUIDAS.—Hardly is any part of Zoology so obscured by absurdities as the history of the Ostrich.

resemble

refemble troops of cavalry, and as fuch have alarmed many caravans. Their life muft be fpent rather hardly in thefe vaft barren folitudes; but there they tafte the fweets of liberty and love. And is not the defert which offers thefe ravifhing pleafures converted into an Elyfian field? To enjoy thefe ineftimable bleffings they fly the prefence of man; but man learns the profit he can derive from them; he haunts them in their moft favage retreats; he feeds on their eggs, their blood, their fat, their flefh; decks himfelf with their plumes; and perhaps he flatters himfelf with the hope of completely fubduing them, and of ranking them among the number of his flaves. The advantages which the domeftication of the Oftrich promifes are fo important, as to threaten its fecurity, even in the deferts.

Whole nations have merited the name of *Struthophagi*, from the cuftom of feeding on the Oftrich; and thefe people bordered on the *Elephantophagi*, who had not better cheer. For this difh Apicius directs, and with great reafon, a poignant fort of fauce; which proves that it was ufed among the Romans; but we have other evidences. The Emperor Heliogabalus once ordered the brains of fix hundred Oftriches to be ferved up for a fingle repaft. That monfter, it is well known, had a whim of eating each day only a fingle kind of food; fuch as pheafants, hogs, pullets, &c. and the Oftrich was of the

number,

number, but feafoned no doubt after the *recipe* of Apicius. Even at prefent the inhabitants of Lybia, Numidia, &c. feed upon tame ones ; eat the flefh, and fell the feathers: yet neither the dogs nor the cats would touch the fragments of the Oftrich diffected by Vallifnieri, though it was frefh and florid. It is indeed true that the Oftrich was extremely lean, and was perhaps old ; but Leo Africanus, who tafted them in their native deferts, informs us, that they were feldom eaten, except when young, and even after being fattened. The Rabbi, David Kimchi, adds, that the females were preferred ; and perhaps the flefh was rendered more palatable by caftration.

Cadamofto and fome other travellers mention their having tafted Oftrich eggs, which they found not to be unpleafant food ; De Brue and Le Maire affirm, that a fingle one is a meal fufficient for eight men ; and others affert, that it weighs as much as thirty hen eggs ; but even this is far fhort of fifteen pounds.

The fhell of thefe eggs is formed into various forts of cups, which in time grow hard, and in fome degree refemble ivory.

When the Arabians have killed an Oftrich, they open its throat, and make a ligature below the incifion ; three or four of them then take it, and fhake it backwards and forwards, as we would rinfe a bottle: the ligature being then removed, a confiderable quantity of *manteca*,

of

of the confiftence of concrete oil, oozes from the
hole. They extract fometimes as much as
twenty pounds from a fingle Oftrich; and
this *manteca* is nothing but the blood of the animal mixed, not with the flefh as has been alleged, fince there is none on the belly and breaft,
but with that fat, which in plump Oftriches
forms, as we have faid, a coat feveral inches
thick on the inteftines. The natives of the country affert that the *manteca* is pleafant to eat,
but occafions a loofenefs.

The Ethiopians flay the Oftriches, and fell
their fkins to the merchants of Alexandria. The
leather is very thick, and the Arabians formerly
made it into inner jackets, which ferved inftead
of a cuirafs and buckler. Belon faw large quan-
tities of thefe fkins ftripped of their feathers in
the fhops at Alexandria.—The long white plumes
of the tail and wings have always been highly
efteemed; the ancients ufed them for decoration,
and appropriated them to diftinguifh the military profeffion, in which they fucceeded to the
feathers of the fwan; for birds have always
furnifhed the polifhed nations, as well as the
favage tribes, with the materials of ornament.
Aldrovandus informs us, that there are ftill pre-
ferved at Rome two ancient ftatues, one of
Minerva and another of Pyrrhus, in which the
helmet is decked with the plumes of the Oftrich.
It feems probable that thefe alfo compofed the
bunch worn by the Roman foldiers, which Po-

lybius

lybius mentions, and which confifted of three black or red feathers, about a cubit in height. In Turkey, even at prefent, a Janiffary, who has diftinguifhed himfelf by his warlike atchievements, is entitled to wear them in his turban; and in the feraglio, the Sultana, when fhe would heighten her charms to obtain a gentler conqueft, employs them to decorate her perfon. In the kingdom of Congo, thefe feathers are mixed with thofe of the peacock, to form enfigns of war; and the ladies of England and of Italy make them into a fort of fans. It is well known what prodigious confumption is made of them in Europe for hats, helmets, theatrical dreffes, furniture, canopies, funeral decorations, and even for female ornaments : and indeed it muft be allowed, that they have a fine effect, both from their natural and their artificial colours, and from their gentle waving motion. But it may be proper to obferve, that the feathers moft admired, are thofe which are plucked from the animal while alive; and are known by this property, that when the quill is preffed by the fingers, it difcharges a bloody liquor, while thofe that are pulled after death are dry, light, and very liable to worms.

The Oftriches, though inhabitants of the deert, are not fo wild as might be fuppofed. All travellers agree in faying, that they are eafily tamed, efpecially when young. The people of Dara, of Lybia, &c. breed them in flocks, and undoubtedly

undoubtedly obtain from them feathers of the best sort, which are only gotten from living Ostriches. They even grow domestic without much trouble, merely from the habit of seeing men, and receiving food, and experiencing kind treatment. Brue, having purchased two of them at Seriupate on the coast of Africa, found them quite tame when he arrived at Fort St. Louis.

They have even been brought farther than domestication, and some have trained them to carry their rider like a horse. Nor is this a modern invention; for the tyrant Firmius, who reigned in Egypt about the end of the third century, used, it is said, to be carried on large Ostriches. Moore, an Englishman, mentions, that he saw at Joar in Africa, a man travelling on an Ostrich. Vallisnieri speaks of a young man who mounted on an Ostrich, exhibited at Venice before the rabble; and Adanson saw, at the factory of Podor, two young Ostriches, the strongest of which ran faster than the best English hunter, though it carried two negroes on its back. All this proves that these animals, though not absolutely intractable, are yet of a stubborn nature, and if they could be taught so much as to keep together in flocks, and return to their stable, and even to allow themselves to be mounted, it would still be difficult and perhaps impossible to instruct them to obey the reins, to feel the wish of the rider, and comply with it.

We

We fee, even from the relation of Adanfon, that the Oftrich of Podor did not make a diftant ftretch, but only took feveral turns round the hamlet, and that its courfe could only be ftopped by throwing fome obftacle in its way. Docile to a certain degree of ftupidity, it feems intractable from its nature; and this muft be really the cafe, fince the Arab, who has tamed the horfe and enflaved the camel, has never completely fubdued the Oftrich; but till this be obtained, advantage can neither be derived from its fpeed nor its force, for the ftrength of an indocile domeftic is always hurtful to its mafter.

But, though the Oftriches run fafter than the horfe, they are yet caught by means of horfes; but to fucceed requires dexterity. The plan which the Arabs take is to keep conftantly within fight of them, without preffing them too hard; they are particularly careful not to fuffer them to feed, though not to difturb them fo much as to tempt them to make their efcape by a fudden flight. And the execution is the more eafy, fince they feldom hold a ftraight courfe, but almoft always defcribe a circle. The Arabs therefore, tracing a fmaller circle within the other, can always keep their proper diftance, and yet pafs over much lefs ground. After a day or two, when the Oftriches are worn out with fatigue and want of food, the horfemen fuddenly dart upon them at full gallop, endeavouring as much as poffible to drive them againft the wind, and

and they kill them with the blows of a ftick, that the blood may not fpoil the fine white of the plumes. It is faid, that when they feel themfelves overcome, and unable to efcape from the hunters, they conceal their head, and imagine that they cannot then be perceived; but that abfurdity muft be afcribed to thofe who attempt to conjecture their intentions; it is evident that they only try to protect that part of their body at laft, which is at once the weakeft and the moft effential.

The *Struthophagi* had another method of catching thefe animals: they covered themfelves with an Oftrich's fkin, and flipping their arms into the neck, they imitated all the ordinary motions of the Oftrich itfelf, and thus were able to get near it and furprife it. In the fame manner, the favages of America difguife themfelves like roebucks, to catch the roe-bucks.

Dogs and nets alfo have been ufed, but it would feem that the horfe is moft commonly employed; and this circumftance alone is fufficient to explain the antipathy which has been fuppofed to fubfift between the horfe and the Oftrich.

In running, it fpreads its wings and the large feathers of its tail, not with the view to affift its motion, as I have already faid, but from the common effect of the correfpondence of mufcles; juft as a man running throws about his arms, or an elephant that turns upon its hunter, erects and

displays

diſplays its large ears. But the complete proof
that the Oſtrich does not raiſe its feathers to in-
creaſe its celerity is, that it ſpreads them, even
when it runs againſt the wind; in which caſe
they can only retard its progreſs. If we conſider that bulk and ſwiftneſs are combined in the
Oſtrich, we muſt be convinced that they are
extremely ſtrong; yet it preſerves the manners
of the granivorous animals: it never attacks the
weak, and ſeldom oppoſes the aſſaults of the
ſtrong. Covered with a hard thick coat of
leather, and furniſhed with a broad *ſternum*,
which ſerves for a breaſt-plate, and defended by
another callous breaſt-plate, it hardly feels the
application of external blows; and it eſcapes
from the greater and more ſerious dangers by
the rapidity of its flight. If it ſometimes makes
reſiſtance, it employs its bill, the points of its
wings, and particularly its feet. Thevenot ſaw
one which overthrew a dog by a blow of its
foot. Belon ſays that it could run down a man;
but that in its flight it throws ſtones at its pur-
ſuer. I doubt the fact, eſpecially as its velocity
would only diminiſh that of the ſtones which it
might throw, the two motions being oppoſite.
Beſides, this fact, advanced by Pliny, and copied
by many others, does not appear to be confirmed
by any modern of credit, and it is known that
Pliny poſſeſſed more genius than diſcernment.

　Leo Africanus ſays, that the Oſtrich wants
the ſenſe of hearing; yet we have already ſeen
　　　　　　　　　　　　　　　　　　　that

that it appears to have all the organs which give thofe fenfations; the aperture of the ears is even very wide, and is not obftructed by feathers. It is probable therefore, that either like the grous, it is only deaf in the feafon of love; or that what has been imputed to its deafnefs, is really the effect of ftupidity.

It is moft likely that this is the feafon when its cry is heard: this happens feldom, for few perfons have mentioned it. The facred writers compare its cry to moaning, and it is even pretended that its Hebrew name *jacnah* is derived from *ianah*, which fignifies *to howl*. Dr. Brown fays, that it refembles the wailing of a hoarfe child, and is ftill more plaintive; how gloomy is it then, and even terrible, to ufe the expreffion of Sandys, to travellers who penetrate with timorous apprehenfions into the immenfity of thefe deferts, where every living being, man not excepted, is an object of dread and danger!

The T O U Y O U.

Struthio Rhea, Linn. Gmel. Borowſk.
Rhea Americana, Lath. Ind.
Rhea, Briſſ.
Struthio Nothus, Klein.
Struthio-camelus Americanus, Ray and Will.
Struthio Emeu, Nieremb.
Nhandaguacu Braſilienſis, Marcg. and Piſo.
The American Oſtrich, Will. Damp. Lath.

WHAT is called the Oſtrich of South Ame‑
rica, or the *Oſtrich of the Straits of Ma‑
gellan and Guiana,* is quite different from the
Oſtrich; and Le Maire is, I believe, the firſt
traveller, who, deceived by ſome traces of re‑
ſemblance to that African bird, has given it the
ſame name. Klein, who perceived that it was
of another ſpecies, is contented with calling it
the Baſtard Oſtrich. Barrere has ſometimes
termed it the *Heron,* ſometimes the *ferrivorous
Crane,* and ſometimes the *Long-necked Emeu.*
Others have with more accuracy applied a com‑
pound name, expreſſive of ſeveral of its quali‑
ties, *the Gray Caſſowary with the Oſtrich-bill.*
Moehring and Briſſon prefer the Latin name
rhea, to which the latter annexes the Ameri‑
can epithet *Touyou,* formed from *Touyouyou,* by
which it is generally known in Guiana. The
ſavages ſettled in other parts of the continent

10   have

have given it different names: *Yardu*, *Yandu*,
Andu, and *Nandu-guacu*, in Brazil * ; *Sallian*
in the ifland of Maragnan † ; *Suri* in Chili,
&c. So many names have been beftowed on
an animal with which we were fo lately made
acquainted! For my part, I fhall readily adopt
that of *Touyou*, which Briffon has applied or
rather retained, and I fhall not hefitate to prefer this barbarous word, which has probably
fome analogy with the voice or cry of that bird,
to the fcientific terms, which only ferve to convey falfe notions, and to new names, which
mark no character, no effential property of the
animal on which they are beftowed.

Briffon feems to imagine that Aldrovandus
meant to figure this bird by the name of *Avis
Eme;* and it is true that we find, in the 541ft
page of vol. iii. of his Ornithology, a plate
which reprefents the Touyou and the Caffowary,
according to the two plates of Nieremberg, and
that it is infcribed in large characters with the
words AVIS EME; in the fame manner as the
figure of the Touyou in Nieremberg bears the
name of *Emeu.* But it is obvious that thefe two
titles have been added by engravers or printers
who were little acquainted with the views of the
authors: for Aldrovandus does not mention a
fingle word of the Touyou, and Nieremberg only

* Nieremberg, Marcgrave, and Pifo.
† Hift. Gen. des Voyages, tome xiv.

calls

calls it *Yardou, Suri,* and the *Occidental Oftrich*;
and both, in their defcription, apply the names
Eme and *Emeu* to the Caffowary of Java alone.
In fhort, to avoid confufion of names, the *Eme*
of Aldrovandus and the *Emeu* of Nieremberg
ought henceforth to be excluded from the fynonyms of the *Touyou.* Marcgrave fays, that
the Portuguefe call it *Ema* in their language;
but the Portuguefe, who had often vifited the
Eaft Indies, were acquainted with the *Emeu* of
Java, and applied that name to the Touyou
of America, which was more analogous to it
than any other bird; for the fame reafon that
we have applied the term *Oftrich* to the fame
Touyou. We muft therefore confider it as an
eftablifhed point, that the *Emeu* belongs exclu-
fively to the Caffowary of the Eaft Indies, and
has no relation to the Touyou, or any other
American bird.

In this detail of the different names of the
Touyou, I have in part pointed out the different
countries where it is found. It is a native of
South America, but is not equally common in
every province of that extenfive country. Marcgrave informs us, that it is feldom feen in the vicinity of Fernambuca, and is equally rare in
Peru, and along the moft populous coafts; but
it is more frequent in Guiana *; in the feigno-
ries of Seregippe and of Rio-grande; in the

* Barrere.

interior

interior provinces of Brazil*; in Chili †; in
the vaſt foreſts ſituated north from the mouth
of the De la Plata ‡; and in the vaſt ſavannas
which ſtretch on the ſouth of that river, and in
all the *terra Magellanica*, as far as Port Deſire,
and even to the coaſt bordering on the ſtraits of
Magellan §. Formerly ſome diſtricts in Para-
guay ſwarmed with theſe birds, eſpecially the
plains watered by the Uraguay; but as the inhabitants multiplied, they killed great numbers,
and forced the reſt to retire. Captain Wood
aſſures us, that though they abound on the
northern coaſt of the Straits of Magellan,
there is not one to be found on the ſouthern
ſhore: and notwithſtanding Coreal's affirming
that he ſaw them on the iſlands in the South Sea,
theſe ſtraits ſeem to bound the climate ſuited
to the Touyou, as the Cape of Good Hope
terminates the region of the Oſtrich; and the
iſlands where Coreal ſaw the Touyous were
probably contiguous to the eaſtern ſhores of
America, beyond the Straits of Magellan. It
alſo appears that the Touyou, which, like the
Oſtrich, delights in the heat of the torrid zone,
is yet not ſo much affected by cold; for the
promontory which runs into the Straits of Magellan is nearer the Pole, than the Cape of Good
Hope, or any other climate, where the Oſtrich

* Marcgrave. † Hiſt. des Gen. Voyages, tome xiv.
‡ Hiſt. des Incas. § Wafer.

has fpontaneoufly fixed its refidence: but, as according to all accounts, the Touyou is alfo, like the Oftrich, entirely a terreftrial bird, and incapable of flying; and as South America is feparated from the ancient continent by immenfe oceans; it would follow that we ought not to expect to find Touyous in our continent, any more than to difcover Oftriches in America: and this inference agrees with the teftimony of travellers.

The Touyou, though fomewhat inferior to the Oftrich, is yet the largeft bird in the New World, the full grown ones being fix feet high *. Wafer, who meafured the thigh of one above the ordinary fize, found it almoft equal to that of a man. It has the long neck, the fmall head, and the flat bill of the Oftrich; but in other refpects, it refembles the Caffowary. I find even in the hiftory of Brazil, written by the Abbe Prevot †, but no where elfe, mention made of a kind of horn which this bird has on its bill, and which, if it really exifted, would be another point of analogy with the Caffowary.

Its body is of an oval fhape, and, when cloth-ed with all its feathers, appears almoft entirely round. Its wings are very fhort, and of no ufe

---

* In the figure that Nieremberg gives, there is a fort of cap on the crown of the head, which refembles the hard callous fkin that, according to Dr. Brown, is found on the fame part of the Oftrich: but no mention is made of this cap in Nieremberg's de-fcription, or in any other.

† Hift. Gen. des Voyages, tome xiv.

in

in flying; though they are faid to affift it in running. It has on the back and about the rump long feathers, which fall back, and conceal the *anus*, and it has no other tail. Its plumage is all gray on the back, and white on the belly. It is very tall, having three toes to each foot, all anterior; for we cannot confider as a toe that callous round tubercle placed behind, and on which the foot refts as on a claw. To this conformation is imputed the difficulty with which it ftands on a fmooth furface, and of walking on it without falling; in compenfation, however, it runs very fwiftly in open ground, raifing fometimes one wing, fometimes another, but for purpofes that have not yet been well afcertained. Marcgrave fuppofes, that it is with the view of fpreading a fort of fail to catch the wind; Nieremberg, that it is to throw the fcent from the dogs which purfue it; Pifo and Klein, that it is to change frequently the direction of its courfe, by thefe zig-zags to avoid the arrows of the favages; and others imagine, that it feeks to increafe its fpeed by pricking itfelf with a kind of needle with which its wings are armed *. But whatever be the intention of the Touyous, certain it is that they run with aftonifhing velocity, and that it is difficult for any hound to overtake

---

* It muft be obferved that neither Pifo, Marcgrave, nor any who ever faw the Touyou, take the leaft notice of this wing-fpur; and it is not unlikely that it was beftowed merely from the analogy fuppofed to fubfift between it and the African Oftrich.—What miftakes are occafioned by the confufion of names?

them.

them. It is related of one, that perceiving itſelf
ſtopt, it darted with ſuch rapidity as diſengaged
itſelf from the dogs, and eſcaped to the mountains*.   It being impoſſible to outſtrip them by
agility, the ſavages are obliged to employ addreſs,
and to lay ſnares for catching them †.   Marcgrave ſays, that they live on fleſh and fruits ;
but if they had been obſerved with attention, it
would undoubtedly have been diſcovered which
of theſe two kinds of aliments they prefer.   For
want of facts, we may conjecture that theſe
birds having the ſame inſtinct with the Oſtriches,
and other frugivorous animals, viz. that of ſwallowing ſtones, iron, and other hard ſubſtances,
that they are alſo frugivorous, and that if they
ſometimes eat fleſh, it is either becauſe they are
preſſed by hunger, or, like the Oſtrich, their
ſenſes of taſte and ſmell being only in an inferior degree, they ſwallow indiſcriminately whatever comes in their way.

Nieremberg relates very ſtrange ſtories in regard to their mode of propagation.   According
to this writer, the male performs the office of
hatching the eggs ; for this purpoſe he aſſembles
twenty or thirty females to lay in the ſame neſt,
and after they have depoſited their eggs, he drives
them forcibly away, and covers them himſelf,
taking the ſingular precaution however to remove two of the eggs from the neſt ; theſe two

* Navigations aux Terres Auſtrales.
† Hiſtoire Gen. des Voyages, tom. xiv.

become

become addle, which the male perceiving, breaks one of them, which invites a multitude of flies, beetles, and other fmall infects, on which the young Touyous feed ; and after the firft is confumed, he opens another for the fame purpofe. But all this may happen, without recurring to an extraordinary fuppofition ; the addle eggs may be crufhed by accident, and infects attracted, which fometimes ferve to nourifh the young Touyous. We can fufpect here the intention of the parent only ; for thefe intentions, which we liberally impute to animals, always form the romance of natural hiftory.

With refpect to the affertion, that the male takes upon himfelf, to the exclufion of the females, the care of hatching, I am much inclined to doubt the fact, conceive it as not authentic, and as inconfiftent with the order of nature. But it is not enough to point out the caufes of error ; we ought, as much as we are able, to difcover the caufes of it, which fometimes lead us alfo to the truth. I fhould therefore imagine that this report is occafioned by the finding of tefticles in fome fitters ; and perhaps an appearance of the *penis*, as is obferved in the female Oftrich, which feemed to evince it to be a male.

Wafer fays, that he faw in a defert tract fituated on the north of the De la Plata, about the thirty-fifth degree of fouth latitude, a number of the eggs of the Touyou in the fand, where, according to him, thefe birds left them to hatch.

If

If this fact be true, the defcription which Nieremberg gives with refpect to the incubation
of thefe eggs, can only refer to a climate that is
colder, and nearer the pole. In fact, the Dutch
found near Port Defire, which is in the forty-
feventh degree of fouth latitude, a Touyou
that was fitting, which they chaced away, and
reckoned nineteen eggs in the neft. It is for
the fame reafon that the Oftriches feldom or
never hatch at all in the torrid zone, but cover
their eggs at the Cape of Good Hope, where the
heat of the climate is inadequate to the effect.

When the young Touyous are newly excluded from the fhell, they are very familiar, and
follow the firft perfon whom they meet *; but
as they grow up, they acquire experience, and
become fhy †. It appears that in general their
flefh is pleafant to eat; though that of the old
ones indeed is tough, and of a bad flavour.
This flefh might be improved by raifing flocks
of young Touyous, which would be eafy, confidering their aptnefs to be tamed; and, by fat-
tening them, and ufing the precautions which
have fucceeded with the turkey, which alfo de-

---

* " I have myfelf," fays Wafer, " been followed by many of
thefe young Oftriches, which are inoffenfive and unfufpicious."
DAMPIER.

† " There is a great number of Oftriches in this ifland of Port
Defire, which are very wild." *Voyage des Hollandois aux Indes Orientales.*—" I faw at Port Defire three Oftriches, though I could not
get near enough to fire at them; as foon as they perceived me they
fled." *Navig. aux Terres Auftrales.*

rives

rives its origin from the warm and temperate countries on the continent of America.

Their feathers are far from being fo beautiful as thofe of the Oftrich ; and Correal even afferts that they are totally ufelefs. It is to be wifhed, that inftead of telling us their little value, travellers had given us a precife idea of their texture.—Too much has been written on the Oftrich, and too little on the Touyou. In giving a hiftory of the former, the chief difficulty confifts in collecting all the facts, in comparing the relations, in difcuffing the opinions, and in feparating the truth from a heap of rubbifh. To difcourfe on the Touyou, we muft often conjecture what *is*, from what *may be ;* we are obliged to comment, to interpret, to fupply ; and when facts cannot be had, to fubftitute what is probable in their ftead ; and, in a word, to remain in fufpence till future obfervations can be procured to fupply the blanks in its hiftory. [A]

[A] The fpecific character of the Touyou, *Struthio Rhea,* LINN. is, that " the foot has three toes, the hind one rounded " LATHAM makes it a genus of which only one fpecies, the *Americana,* is known. " The *bill* is ftraight, depreffed, and roundifh at the tip ; the *wings* are unfit for flying ; the lower part of the *thighs* naked ; the *feet* are furnifhed with three toes before, and behind with a round callous bump."

The GALEATED CASSOWARY.

Le Casoar, Buff.
Struthio Cassuarius, Linn. and Gmel.
The Cassowary, or *Emeu,* Will. Ray, Briss Klein, &c.

THE Dutch are the first who shewed this
bird in Europe; they brought it, in 1597,
from the island of Java, on their return from
the first voyage which they performed to the
East Indies *. The natives of the country call
it *Eme,* from which the French have formed the
word *Emeu.* It was also named *Cassowary,* which
I have adopted, since it has been appropriated to
this bird.

The Galeated Cassowary, though not so large
as the Ostrich, is apparently more bulky; because its mass is nearly the same, and its neck
and feet are shorter and thicker in proportion;
and its body is more protuberant, which gives it
a heavier look.

The one described by the Academicians was
five feet and a half long, from the point of the
bill to the extremity of the claws; that observed by Clusius was a fourth smaller. Houtman
reckons it double the bulk of the swan; and
other Dutchmen mention it as of the size of a

* *Hist. Gen. des Voyages, tome* viii, and CLUSIUS, *Exotic.*

sheen,

sheep. This variety of measures, so far from affecting the truth, is what alone informs us of the real magnitude of the Galeated Cassowary; for the size of an individual is not that of the species, and to estimate that properly, we must consider it as a quantity varying between certain limits. Hence a naturalist who compares with judgment the descriptions of different observers, will have more precise and accurate ideas of the species, than an observer who is only acquainted with a single individual.

What is most remarkable in the figure of the Galeated Cassowary, is that kind of conical helmet, the fore-part of which is black, and the rest yellow, which covers the face from the bottom of the bill to the middle of the crown of the head, and sometimes stretches farther. This helmet is formed by the protuberance of the bones in that part of the cranium, and is sheathed by a hard covering, consisting of several concentric plates analogous to the substance of an ox's horn. Its entire shape resembles a truncated cone, three inches high, an inch diameter at the base, and three lines at its vertex. Clusius thought that this helmet dropped every year with the feathers, in the season of moulting; but the Academicians have properly observed, that the external sheath only could thus fall, and not the inner substance, which, as we have said, forms a part of the bones of the skull; and they even add, that during the four years that

13   this

this bird was kept in the *menagerie* at Verſailles, they could not perceive that this ſheath was ever detached. However, this might have happened through length of time, and by a kind of ſucceſſive exfoliation, as in the bill of many birds ; and this procefs might have efcaped the obſervation of the keepers of the *menagerie.*

The iris is of a topaz-yellow, and the cornea is remarkably fmall, compared with the ball of the eye*, which gives the animal a ſtrange wild appearance ; the lower eye-lid is the largeſt, and the upper is fet in the middle with a row of fmall black hairs, which form an arch over the eye like the brow, and this, together with the opening of the bill, produces a threatening afpeƈt. The exterior orifices of the noſtrils are feated very near the point of the upper bill.

In the bill we muſt diſtinguiſh the materials which ſerve to cover it. They are three ſolid pieces, two of which form the circumference, and the third conſtitutes the upper ridge, which is much more elevated than in the Oſtrich ; the three are ſheathed with a membrane which fills up the interſtices.

The upper and lower mandibles of the bill have their edges a little furrowed near the end, and feem each of them to have three points.

---

* The ball of the eye was one inch and a half in diameter ; the cryſtalline lens four lines, and the cornea only three lines. *Mem. pour ſervir a l'Hiſt. des Anim.*

The

The head and the arch of the neck are
sprinkled with a few small feathers, or rather
with some black straggling hairs; so that on
these parts the skin appears bare. The colours
and their dispositions are various; commonly
blue on the sides, violet under the throat, red
behind in many parts, but especially in the
middle of the neck; and these red parts are
more prominent than the rest, on account of
wrinkles, or oblique furrows.

The holes of the ears were very large in the
Galeated Cassowary described by the Academicians; very small in the one described by Clusius; but in both they were disclosed, and beset
like the eyelids with small black hairs.

Near the middle of the fore-part of the neck,
and where the great feathers have their origin,
rise two barbels which are red and blue, and
round at the ends, and which Bontius places in
his figure immediately above the bill, as in poultry. Frisch delineates four; two long ones on
the sides of the neck, and two before that are
smaller and shorter: the helmet also appears
larger in his figure, and approaches the shape
of a turban. There is in the king's cabinet a
head which seems to be that of the Galeated
Cassowary, but which has a tubercle different
from what is ordinary. It will require time and
observation to ascertain whether these varieties,
and those which we shall afterwards mention,
be constant or not; if some of them be not
owing

owing to the inaccuracy of the defigners, or are only fexual differences. Frifch pretends that he difcovered in two ftuffed Galeated Caffowaries, the diftinguifhing marks between the males and females ; but he does not inform us in what thefe confift.

The wings of the Galeated Caffowary are ftill fmaller than thofe of the Oftrich, and equally unfit for flying. They are armed with points, and thefe are even more numerous than in the Oftrich. Clufius found four or five of them ; the Academicians five; and in Frifch's figure there are evidently feven : thefe are like the pipes of feathers, and appear red at the end, and are hollow through their whole extent. They contain within their cavity a fort of marrow fimilar to what is found in the fprouting feathers of other birds. The middle one is near a foot in length, and about three lines in diameter, it being the longeft of all : thofe placed on either fide diminifh gradually like the fingers of the hand, and nearly in the fame order. Swammerdam ufed them inftead of a pipe to inflate very delicate veffels, fuch as the *trachea* of infects, &c. It has been faid, that the wings of the Caffowary were intended to accelerate its motion ; others have conjectured that they only ferved them like fwitches to affift them in ftriking ; but no one can affert that he ever faw what ufe the bird really makes of them. The Caffowary has alfo another property common to the

the Oftrich, viz. it has but one kind of feathers
over its whole body, wings, rump, &c. though
moft of thefe feathers are double, each root
fending off two branches of different lengths:
nor is the ftructure uniform throughout; the
branches being flat, black, and fhining, divided
underneath into knots, each of which produces
a beard or thread, with this difference, that from
the root to the middle of the branch, thefe
threads are fhorter, more pliant, ramify more,
and are covered with a kind of tawny down;
whereas, from the middle of the fame branch
to its extremity, they are longer, harder, and
of a black colour, and as thefe laft cover the
others, and are the only ones that appear, the
Caffowary feen at a diftance refembles an
animal clothed with hair, like bears or wild
boars. The fhorteft feathers are on the neck,
the longeft round the rump, and the middle
fized on the intermediate fpace. Thofe of the
rump are fourteen inches long, and hanging
over the hinder part of the body, they fupply
the place of the tail, which is totally wanting.

It has, like the Oftrich, a naked and callous
fpace on the *fternum*, where the weight of the
body refts when the bird fits; and this part is
ftill more prominent in the Galeated Caffowary
than in the Oftrich *.

The thighs and legs are clothed with feathers
almoft to the knees, and thefe feathers were of

* Voyages de la Compagnie Hollandoife.

an

an afh-gray in the fubject which Clufius examined; the feet, which are thick and ftout,
have three toes, and not four, as Bontius affirms;
all of them directed forwards.   The Dutch relate, that the Galeated Caffowary employs its
feet for defence; ftriking backwards like a horfe,
according to fome; and according to others, darting forwards againft the affailant, it throws him
back with its feet, and ftrikes his breaft with violent blows.   Clufius, who faw one alive in the
gardens of Count Solms at the Hague, fays, that
it makes no ufe of its bill for protection, but
that it attacks its antagonift fideways, by kick-
ing; he adds, that this Count fhewed him a tree
about the thicknefs of his thigh which this bird
had fpoiled, having ftripped off the bark entirely
with its feet and nails.   The Caffowaries kept
in the *menagerie* at Verfailles have not indeed
been obferved to be fo ftrong or fo mifchievous,
but perhaps they were grown tamer than that of
Clufius: befides, they lived in abundance and
in clofer captivity; circumftances which in time
meliorate the difpofitions of fuch animals as are
not altogether wild, enervate their courage, blunt
their original inftincts, and render it impoffible
to diftinguifh thefe from their acquired habits.

The claws of the Caffowary are very hard,
black on the outfide, and white on the infide.
Linnæus fays, that they ftrike with the middle
claw, which is the largeft; yet the defcriptions
and figures of the Academicians and of Briffon
reprefent

reprefent the inner claw as the largeft, which is really the cafe.

Its gait is fingular ; it appears to kick behind, at the fame time it makes a kind of leap forwards. But however ungraceful its motion may be, it is fwifter, we are told, than the beft runner : indeed celerity of motion is fo peculiarly the property of birds, that the tardieft of that tribe excel in the rapidity of their courfe the moft agile of the land animals.

The Caffowary has the tongue indented along the edges, and fo fhort, that it has been faid of it, as the moor cock, that it has none. The one obferved by Perrault was only an inch long, and eight lines broad *. It fwallows any thing that is thrown to it ; that is, any fubftance which its bill will admit. Frifch juftly confiders this inftinct as indicating an analogy to the gallinaceous tribe, which fwallow their aliments entire, without bruifing them with their bills ; but the Dutch, who feem to have wifhed to make the hiftory of this fingular bird ftill more extraordinary by the addition of the marvellous, have not hefitated to affert, that it fwallows ftones, bits of iron, glafs, &c. and even burning coals, without fuffering inconvenience †.

It is alfo faid they eject very foon what they have taken, and fometimes difcharge apples as

* Mem. pour fervir à l'Hiftoire des Animaux.
† Hift. Gen des Voyages, tome viii.

large

large as the hand, and in the fame ftate in which they were fwallowed. Indeed, the inteftinal canal is fo fhort, that the aliments muft foon pafs through it; and fuch as, by their hardnefs, might occafion fome refiftance, muft undergo little alteration in fo fmall a defcent, particularly when the functions of the ftomach are deranged by any difeafe. Clufius was affured, that in thefe cafes, they fometimes ejected hen-eggs, which they are fond of, and quite entire with their fhell; but, on fwallowing them a fecond time, they completely digefted them*. The principal food of this bird, which was the Caffowary belonging to Count Solms, was white bread cut into fmall bits, which proves that it is frugivorous, or rather omnivorous, fince it really eats whatever is offered it, and has the craw and the double ftomach of the animals that live on vegetable fubftances, and at the fame time it has the fhort inteftines of fuch as feed on flefh. The inteftinal canal of the one diffected by the Academicians was four feet eight inches long, and two inches diameter through its whole extent. The *cæcum* was double, and only one line in diameter, and three, four, or five inches long. From this account it appears, that the inteftines of the Galeated Caffowary are thirteen times fhorter than thofe of the Oftrich; and for this reafon, it muft be ftill more voracious, and ftill more difpofed to animal food, which could be

* Clufius. *Exotic.*

afcertained

afcertained if obfervers, inftead of refting fatif-
fied with examining the dead bodies, would
ftudy the habits of the bird while alive.

The Caffowary has a gall-bladder; and its
duct, which croffes the hepatic, terminates higher
than that in the *duodenum*, and the pancreatic
duct is inferted above the cyftic; a conformation of parts quite different from what obtains
in the Oftrich. The organs of generation in
the male are not fo diffimilar: the *penis* rifes
from the upper part of the *rectum*; its form is
that of a triangular pyramid, two inches broad
at the bafe, and two lines at the *apex*; it confifts of two folid cartilaginous ligaments, connected clofely to each other above, but parted
below, and leaving between them a half-channel
covered with fkin. The *vafa deferentia* and
the ureters have no apparent communication
with the perforation of the *penis*; fo that this
part, which feems to fill four principal offices in
the quadrupeds, that of carrying off the urine,
that of conveying the feminal fluid to the female womb, that of contributing by its fenfibility to the emiffion, that of ftimulating the female to melt in the embrace, feems in the Caffowary and the Oftrich to be confined to the
two laft, which are calculated to excite in the
two fexes the neceffary correfpondence of motion in the venereal act.

Clufius was informed that, when the animal
is living, the *penis* fometimes is obferved to

projeƈt from the *anus;* another point of analogy with the Oftrich.

The eggs are of an afh-gray, verging on
greenifh, not fo thick, but longer than thofe of
the Oftrich, fprinkled with a multitude of fmall
tubercles of a deep green; the fhell is not very
thick according to Clufius, who faw feveral of
them; the largeft of all thofe which had fallen
under his notice was fifteen inches round one
way, and a little more than twelve the other *.

The Caffowary has the lungs and the ten air
cells as in other birds, particularly thofe of the
large kind; it has that fcreen or black membrane
peculiar to the eyes of birds, and that inner eye-
lid, which, as it is well known, is attached to
the large angle of the eye by two common
mufcles †, and which is at momentary intervals
drawn back over the *cornea,* by the aƈtion of a
kind of mufcular pulley, which merits all the
curiofity of anatomifts ‡.

The middle of the eaftern part of Afia feems
to be the true climate of the Caffowary, and its
territory begins where that of the Oftrich ends.
The latter feldom paffes beyond the Ganges, as
we have already feen; but the former is found
in the Molucca iflands, and in thofe of Banda,
Java, Sumatra, and the correfponding traƈts on

* "Eggs with excavated points," is Linnæus's expreffion, which
is totally different from what Clufius afferts.
† Hift. de l'Acad. Royal des Sciences, tome ii.
‡ Mem. pour fervir à l'Hift. des Anim.

the

the continent *. It is however far from being
fo numerous as the Oſtrich, ſince a king of Joar-
dam in the iſland of Java preſented Scellinger,
the captain of a Dutch veſſel, with a Caſſowary
as a rare bird. The reaſon probably is, becauſe
the Eaſt Indies are much more populous than
Africa; and it is well aſcertained, that as men
multiply, the wild animals gradually diminiſh,
or retire into the more ſolitary tracts.

It is ſingular, that the Caſſowary, the Oſtrich,
and the Touyou, which are the three largeſt
birds that are known, are all natives of the
torrid zone, which they ſeem to ſhare among
themſelves, each enjoying its own territory,
without incroaching on that of another. They
are really all of them land animals, incapable
of flying, but running with aſtoniſhing ſwiftneſs; all ſwallowing whatever comes in their
way, grain, graſs, fleſh, bones, ſtones, flints, iron,
glaſs, &c. In all, the neck is of great length,
the legs tall and very ſtrong, the claws fewer
than in moſt birds, and in the Oſtrich, there
are ſtill fewer than in the other two; in all, there
is only one ſort of feathers, unlike thoſe of other
birds and different in each of the three kinds;
in all, the head and the arch of the neck are bare,
the tail, properly ſo called, is wanting, the wings
are but imperfect, furniſhed with a few pipes,
without any vanes, as the quadrupeds that in-

† Voyages des Hollandois.

habit

habit the warm countries have lefs hair than thofe of the regions of the north. All of them, in a word, feem to be natural productions of the torrid zone. But notwithftanding thefe points of agreement, they are ftill marked by characters that diftinctly feparate the fpecies; the Oftrich is removed from the Galeated Caffowary and the Touyou, by its fize, by its feet, like thofe of the camel, and by the nature of its plumage : it differs from the Caffowary particularly by its naked thighs and flanks, by the length and capacity of its inteftines, and becaufe it has no gall-bladder; and the Galeated Caffowary differs from the Touyou and the Oftrich, by its thighs being clothed with feathers, almoft to the *tarfus*, by red barbils which hang from the neck, and alfo by the helmet on its head.

But in this laft diftinctive character we ftill perceive an analogy with the other two kinds; for this helmet is nothing but a protuberance of the bones of the cranium, which is covered with a fheath of horn; and we have feen, in the hiftory of the Oftrich and the Touyou, that the upper part of the cranium of thefe two animals was fimilarly defended by a hard callous plate. [A]

[A] The fpecific character of the Galeated Caffowary in the Linnæan fyftem is, that " it has three toes, with the helmet and " barbils naked." Latham erects the Caffowary into a new genus, diftinguifhed by thefe properties: the *bill* depreffed, ftraight, and fomewhat conical; the noftrils oval; the *wings* very fhort, and unadapted

adapted for flying; the lower part of the *thighs* naked; the *feet* consisting of three toes, all of them turned forwards; the *tail* wanting. The Galeated Cassowary has the epithet of *Emeu*, and is discriminated by being " black, its top helmeted, its body beset with " shaggy feathers, its head and top of the neck bare."

A new species of Cassowary has lately been discovered at Botany Bay, and termed the *New Holland* Cassowary. It is much larger than the former, being seven feet two inches long, while the other is only five feet and a half. It runs exceeding swiftly, and its flesh is palatable food. Latham thus characterizes it:—" It is " blackish, its crown flat, its body bristly, its head and neck planted " with quills, its legs serrated behind."—The bill is black; the head, the neck, and the whole of the body, covered with quills, which are variegated with dusky and gray colours; the throat is bare and bluish; the quills on the body are a little bent at the point; the feet are dusky, and behind they are rough through their whole length with protuberances.

The HOODED DODO.

Le Dronte, Buff.
Didus Ineptus, Gmel. and Lath.
Struthio Cucullatus, Linn. 10th edit.
Raphus, Briff.
Cygnus Cucullatus, Nierem. Ray and Will.
Gailus Gallinaceus Peregrinus, Cluf.
Dod-Aerfen, or *Walgh-Vogel,* Herb.

AGILITY is commonly conceived to be peculiarly the property of the winged tribe;
but if we regard it as an effential character, the
Dodo muft be excluded from the clafs; for its
proportions and its movements give an idea of
the moft heavy and awkward of organized beings. Figure to yourfelf a body that is bulky,
and almoft cubical, fupported with difficulty on
two exceedingly thick and fhort pillars, and carrying a head fo ftrangely fhaped, that we might
take it for the whim of a caricature painter;
and this head, refting on a huge fwelling neck,
confifts almoft entirely of an enormous beak, in
which are fet two large black eyes encircled
with a ring of white, and where the parting of
the mandibles runs beyond the eyes, and almoft
quite to the ears; thefe two mandibles, concave
in the middle, inflated at both ends, and bent
backwards at the point, refemble two fharp
fpoons laid on each other, their convexity being
turned outwards: all which produces a ftupid

vora-

voracious appearance, and which, to complete
the deformity, is furnifhed with an edging of feathers, which, accompanying the curvature of the
bafe of the bill, ftretch to a point on the forehead, and then arch round the face like a cowl,
whence the bird has received the name of
Capuchined Swan (Cygnus Cucullatus).

Magnitude, which in animals implies ftrength,
produces nothing in this bird but oppreffive
weight. The Oftrich, the Touyou, and the
Galeated Caffowary, indeed, are alfo incapable
of flying, but they run with aftonifhing fpeed.
The Dodo feems to be clogged by its unwieldy
carcafs, and can hardly collect force fufficient to
drag it along. It is the moft inactive of the feathered race. It confifts, we might fay, of brute
paffive matter, where the living organic particles
are too fparingly diffeminated. It has wings;
but thefe are too fhort and too feeble to raife it
from the ground. It has a tail, but it is difpro-
tioned, and out of place. We might take it for
a tortoife difguifed in the clothing of the winged
tribe; and Nature, in beftowing thefe ufelefs ornaments, feems to have defired to add clumfinefs
to its unwieldy mafs, and to render it more difgufting, by reminding us at the fame time that
it is a bird.

The firft Dutch that faw it in the ifland of
Mauritius, now the Ifle of France *, named it

---

* The Portuguefe had before called that ifland *Ilha do Cirne*;
that is, *Ifland of Swans*; probably becaufe of the Hooded Dodos
they had feen on it, and which they had miftaken for fwans.

Walgh-

Walgh-Vogel, Difgufting Bird, both on account of its ugly figure, and the rank fmell. This fingular bird is very large, and is only inferior in fize to the three preceding; for it exceeds the turkey and the fwan.

Briffon affigns as one of the characters, its having the lower part of the legs naked; yet in the 294th plate of Edwards it is reprefented feathered, not only as low as the leg, but even to the articulation with the tarfus. The upper mandible is blackifh throughout, except at the hook, where there is a red fpot; the holes of the noftrils are placed very near its middle, and clofe to the two tranfverfe folds, which rife at this part on the furface.

The feathers of the Dodo are in general very foft, and their predominating colour is gray, which is deeper on all the upper part of the body and the lower part of the legs, but brighter on the ftomach, the belly, and the whole of the under part of the body. There is fome yellow and white on the quill-feathers of the wings, and thofe of the tail, which appear frizzled, and are but few in number. Clufius reckons only four or five.

The feet and toes are yellow, and the nails black; each foot has four toes, three of which are placed before, and the fourth behind, and this hind one has the longeft nail.

Some have pretended that there was commonly lodged in the ftomach of the Dodo a

ftone

THE HOODED DODO. 393

ftone of the fize of the hand, and to which they
failed not to afcribe the fame origin and the
fame virtues as to the bezoars. But Clufius,
who faw two of thefe ftones of different fhapes,
and bulky, is of opinion, that the bird had fwallowed them like the granivorous clafs, and that
they were not formed in its ftomach.

The Dodo is a native of the iflands of France
and Bourbon, and is probably found alfo on the
neareft parts of the continent, though I know of
no traveller who mentions his feeing it, except
on thefe iflands.

Some Dutch call it *Dodaers;* and the Portuguefe and Englifh, *Dodo;* however, it is named
by the natives *Dronte.* It has alfo been called
Hooded Dodo, Foreign Cock, Walgh-Vogel; and
Mæhring, who has found none of thefe names
to his liking, has formed that of *Ruphus,* which
Briffon has adopted for his Latin defignation, as
if there was any advantage in giving the fame
animal a different appellation in each language,
when the real effect of the multitude of fynonyms is to occafion embarraffment and confufion. " Do not multiply exiftences," was once
the maxim of philofophers; but at prefent we
have conftantly reafon to remind naturalifts not
to multiply names without neceffity. [A]

[A] Linnæus makes the Dodo generic, and to include three
fpecies, viz. The Hooded Dodo of this article, the Solitary Bird,
and the Nazarene Bird. The two latter are joined together in the
following article. The Hooded Dodo has the epithet *Ineptus,* and
is characterized by being " black with whitifh clouds, and its feet
" having four toes."

The SOLITARY DODO, and NAZARENE DODO.

Le Solitaire, et *L'Oiseau de Nazare,* Buff.
Didus Solitarius, et *Didus Nazarenus,* Gmel.

THE Solitary Bird mentioned by Leguat [*]
and Carré [†], and the Bird of Nazareth by
Father Cauche [‡], seem to bear a great resemblance to the Dodo, though they still differ in
several points. I have thought proper to produce what these travellers relate on this subject,
since, if these three names are applicable only to
the same individual species, the different relations will serve to complete the history of the
bird ; if on the contrary, they refer to three different species, what I shall give will be considered as the beginning of the history of each, or at
least as an intimation of a new species to be
examined, in the same manner as it is usual in
geographical charts to mark countries unexplor-
ed. At all events, it is to be desired that those
naturalists, who have an opportunity of examining these birds more closely, would compare
them if possible, and obtain a more precise and

[*] Voyage en Deux Iles Desertes des Indes Orientales.
[†] Hist. Gen. des Voyag.
[‡] Description de l'Ile de Madagascar.

distinct

diſtinct information. Queries alone, made with reſpect to facts with which we are unacquainted, have more than once led to a diſcovery.

The Solitary Dodo of the iſland of Rodrigue is a very large bird, ſince ſome males weigh forty-five pounds. The plumage of theſe is commonly mixed with gray and brown ; but in the females, ſometimes brown, ſometimes a light yellow, predominates. Carré ſays, that the colour of the plumage of theſe birds is gloſſy, bordering on yellow ; he adds, that it is exceedingly beautiful.

The females have a protuberance over the bill reſembling a widow's peak; their feathers bunch out on both ſides of the breaſt into two white tufts, ſomewhat like a woman's boſom. The feathers of the thighs are rounded towards the end in the ſhape of ſhells, which has a very fine effect ; and, as if the females were conſcious of their beauty, they take great pains in arranging their plumage, ſmoothing it with their bill, and adjuſting it almoſt continually, ſo that not a ſingle feather is miſplaced. According to Leguat, their whole appearance is noble and graceful ; and this traveller even affirms that their pleaſing demeanour has often been the means of ſaving their life. If this be the caſe, and if the Solitary and the Dodo be of the ſame ſpecies, we muſt admit a very wide difference between the male and the female in regard to their figure.

This

This bird has some resemblance to the turkey; its legs differ only in being taller, and the bill in being more hooked; its neck is also proportionally longer, the eye black and lively, the head without a crest or tuft, and with scarcely any tail; its hind part, which is round like the buttocks of a horse, is covered with broad feathers.

The wings of the Solitary Dodo do not enable it to fly; but they are not useless in other respects. The pinion-bone swells near the end into a spherical button, which is concealed under the feathers, and serves two purposes; in the first place for defence, to which the bill is also subservient; in the second, to make a kind of clapping or whirling twenty or thirty times on the same side in the space of four or five minutes. In this way, it is said, the male invites his mate with a noise like that of a kestrel, and which is heard at the distance of two hundred paces.

These birds are rarely seen in flocks, though the species is pretty numerous; some affirm even that scarcely two are ever found together *.

They seek unfrequented spots where to lay their eggs; they construct their nest with the leaves of the palm-tree heaped up a foot and a half high; into this nest the female drops an egg much larger than that of a goose; and the male participates in the office of hatching.

* Hist. Gen. des Voyages, tome ix.

During

During the whole time of the incubation, and even that of the education, they fuffer no bird of the fame kind to approach within two hundred paces; and it is pretended that the male drives away the males, and that the female drives away the females; an obfervation which could hardly be made on a bird that paffes its life in the wildeft and the moft fequeftered fpots.

The egg (for it feems that thefe birds lay only one, or rather only cover one at a time) requires feven weeks * to hatch, and the young one cannot provide for itfelf until fome months afterwards. During all that time it is watched with paternal care, and this circumftance alone gives greater force to the inftinctive affection than in the Oftrich, which is abandoned from its birth, and never afterwards receives the foftering affiduities of its parents, and, being without any intimacy with them, is deprived of the advantages of their fociety, which, as I have elfewhere remarked, is the firft education of animals, and which moft of all contributes to develope their native powers; and hence the Oftrich is confidered as the moft ftupid of the feathered creation.

After the education of the young Solitary Dodo is completed, the parents ftill continue

---

* Ariftotle allows thirty days for the incubation of the large birds, fuch as the eagle, the bullard, and the goofe; he does not indeed mention the Oftrich in that place. *Hift Anim.* lib. vi.

united, and on the whole faithful to each other, though fometimes they intermix with other birds of the fame fpecies. The care which in common they have beftowed on the fruit of their union feems to rivet their attachment, and when the feafon again invites, they re- new their loves.

It is afferted, that whatever be their age, a ftone is always found in their gizzard, as in the Hooded Dodo : this ftone is as large as a hen's egg, flat on the one fide and convex on the other, fomewhat rough, and fo hard as to be fit for a whetftone. It is added, that it is always alone in the ftomach, and is too bulky to pafs through the intermediate duct which forms the only communication between the craw and the giz- zard ; and hence it is inferred, that this ftone is formed naturally in the gizzard of the Solitary, and in the fame way as the bezoars. But for my part, I fhould only conclude that this bird is granivorous, and fwallows ftones and pebbles like all the reft of that clafs, particularly the Oftrich, the Touyou, the Caffowary, the Hooded Dodo, and that the paffage between the craw and the gizzard admits of a greater dila- tation than Leguat fuppofed.

The epithet of *Solitary* alone indicates fuf- ficiently its native wildnefs; and this is in- deed what we fhould expect. Bred fequeftered without a fingle companion, deprived of the fociety of its equals, and connected to its parents

only

only by the ties of dependence and want, its latent powers are never awakened and expanded. But it appears ftill more timid than favage; it even ventures to come nigh one, and with an air of familiarity, efpecially if it has little experience, and is not fcared by a fudden onfet; but it can never be tamed. It is difficult to enfnare it in the woods, where it can elude the fportfman by cunning and dexterity in concealing itfelf; but as it does not run faft, it is eafily caught in the plains and open fields; when overtaken, it utters not a complaint, but waftes its grief in tears, and obftinately refufes every kind of food. M. Caron, director of the French Eaft India Company's affairs at Madagafcar, put two of them, from the ifle of Bourbon, on board a veffel, to be prefented to the Royal Cabinet, but they would neither eat nor drink, and died in the paffage.

The proper feafon for catching them is from March to September, which is the winter in thofe countries they inhabit; it is alfo the time when they are fatteft. Their flefh, efpecially when young, is of an excellent flavour.

Such is the general idea which Leguat gives of the Hermit or Solitary Dodo; and he fpeaks not only as an eye-witnefs, but as an obferver, who had for a long time ftudied the habits of the bird; and, indeed, his account, though mar-
red

red in fome places with fabulous notions*, contains more hiftorical details in regard to the Hermit than I have been able to difcover in a crowd
of writings on thofe birds that are more generally and more anciently known.   The Oftrich
has been a fubject of difcourfe for thirty cen-
turies, and yet we are ftill ignorant how many
eggs it lays, and how long its incubation lafts. [A]

The Bird of Nazareth †, fo called, no doubt,
by corruption, becaufe it was found in the ifland
of Nazare, was obferved by F. Cauche in the
ifland of Mauritius.   It is a very large bird,
and more bulky than the fwan.   Inftead of
plumage, its body is entirely covered with a
black down ; yet it has fome feathers, which are
black on the wings and frizzled on the rump,
which ferves for a tail ; it has a thick bill, incurvated fomewhat below ; the legs tall and
covered with fcales, three toes on each foot ; its
cry refembles that of a gofling, and its flefh has
a tolerable relifh.

The female lays only one egg, which is white,
and about the fize of a halfpenny roll.   Befide

* For inftance, he fancies a fort of marriage ceremony is performed at the firft congrefs of the young Hermits ; the ftory of
the ftone in the ftomach, &c.

[A] The fpecific character of the Solitary Dodo : " It is varie-
" gated with gray and dufky, and its feet are furnifhed with four
" toes."

† The ifland of Nazare is of a higher latitude than the ifland
of Mauritius, being feventeen degrees fouth. . . *Defcription de
Madagafcar, par Fr. Cauche.*

it

it, there is generally found a white ftone of the fize of a hen's egg; and this perhaps ferves the fame purpofe with the balls of chalk which the farmers place in the nefts where they wifh their hens to lay. The Nazare depofits its egg on the ground in the forefts, on fmall heaps of grafs and leaves which it makes. When the young one is killed, a gray ftone is found in its gizzard. The figure of this bird, it would appear from a note, is to be met with in " *the Jour-* " *nal of the fecond Voyage performed by the Dutch* " *to the Eaft Indies;*" and they called it the *Bird of Naufea.* Thefe laft words feem decidedly to afcertain the identity between the fpecies of this bird and that of the Dodo; and would indeed amount to a proof, if their defcriptions did not mark effential differences, particularly in the number of their toes. But not to enter into a minute difcuffion, or venture to folve a problem, for which we are not in poffeffion yet of the neceffary *data*, I fhall barely ftate thofe points of refemblance and contraft, which may be difcovered from a comparifon of the three defcriptions. [A]

It readily appears then from a comparifon, that thefe three birds belong to the fame climate, and are natives of almoft the fame tracts.

[A] Gmelin and Latham beftow on this bird the appellation *Didus Nazarenus,* or *Nazarene Dodo.* They regard it as a diftinct fpecies from the Solitary Dodo, and as difcriminated by being black, and having three toes on each foot.

The

The Hooded Dodo inhabits the Iflands of Bourbon and the Ifle of France ; the Hermit refided
in the ifland of Rodrigue, when it was a mere
wafte, and has been feen in the Ifland of Bourbon ;
the Bird of Nazare has been found in that ifland
and in the Ifle of France : but thefe four iflands
are contiguous to each other ; and it is to be remarked, that none of the birds has ever been difcovered on the continent.

All thefe birds refemble each other more or
lefs in point of fize, inability to fly, the form of
their wings, of their tail, and their whole body ;
and in all of them, one or more ftones have
been found in their gizzard, which implies that
they are granivorous.   In all of them, the gait
is flow ; for though Leguat does not mention
that of the Hermit, we can eafily infer from
the figure which he gives of the female, that it
is a fluggifh bird.

Finally, Comparing them two and two, we
perceive that the plumage of the Hooded Dodo
approaches that of the Hermit in its colour, and
that of the Bird of Nazare, by its downy quality ; and that thefe two laft agree alfo in only
laying and hatching a fingle egg.

Both the Dodo and the Bird of Nazare have
been confidered as having a difgufting appearance.

Such are the refemblances.—The differences
are as follow :—

The Hermit has the feathers on its thighs
rounded at the end like fhells ; which proves

that

that they are true feathers, such as thofe of ordinary birds, and not a kind of down, as is the
cafe with the Hooded Dodo and the Bird of
Nazare.

The female Hermit has two white tufts of
feathers on its breaft; nothing fimilar is mentioned in regard to the female of the two
others.

In the Hooded Dodo, the feathers which border the bafe of the bill are difpofed in the fhape
of a cowl; and the appearance is fo ftriking,
that it has given foundation for its characteriftic
name (*Cycnus Cucullatus*). Befides, the eyes
are placed in the bill, which is no lefs remarkable; and we cannot doubt that Leguat faw
nothing like this in the Hermit, fince he only
mentions with regard to that bird, which he
had viewed fo often, that there is neither creft
nor tuft on its head; and Cauche, in fpeaking
of the Bird of Nazare, takes no notice of any
thing of this kind.

The two laft are tall; but the Hooded Dodo
has very thick fhort legs.

The Hooded Dodo and the Hermit, whofe
legs are faid to refemble thofe of the turkey, have
four toes, and the Bird of Nazare, according to
Cauche, has only three.

The Hermit makes a remarkable beating with
its wings, which has not been obferved in the
others.

Laftly, It appears that the flefh of the Hermits, and efpecially of the young ones, is ex-

cellent;

cellent; that of the Bird of Nazare indifferent,
and that of the Hooded Dodo, bad.

If this comparifon, which has been made
with the greateft accuracy, does not allow us
to decide on the queftion propofed, it is becaufe
thefe obfervations are neither fufficiently numerous nor certain. It is therefore to be wifhed, that thofe travellers, and particularly thofe
naturalifts, who have it in their power, would
examine thefe three birds, and form an exact
defcription of them, attending chiefly to the
following points :

The fhape of the head and bill.

The quality of the plumage.

The form and dimenfions of the feet.

The diftinguifhing marks between the male
and female.

The differences between the chicks and adults.

Their manner of walking and running.

Adding as much as poffible of what can be
learnt from the natives refpecting their pairing,
copulating, building their neft, and hatching.

The number, fhape, colour, weight, and bulk
of their eggs.

The time of incubation.

The manner of rearing their young.

Their mode of feeding.

Finally, The form and dimenfions of their
ftomach, of their inteftines, and of their fexual
organs.

NOTES

N O T E S

By *the* TRANSLATOR.

Vol. I. Page 12. Line 13. [*Denſity of the Air.*]

THE obſervation of our ingenious author, that
ſound is much more audible during the night
than in the heat of the day, is curious and intereſting;
but the reaſon which he aſſigns for it, though ſpecious,
appears to be altogether inadequate to the effect. Air
expands one-four-hundredth of its bulk for every degree of heat on Fahrenheit's ſcale; and therefore, ſup-
poſing the difference of temperature between the night
and the day to be twenty degrees, which is a very large
allowance, the variation of the air's denſity would not
exceed one-twentieth; a quantity too ſmall ſurely to
be diſtinctly perceived. The atmoſphere often under-
goes greater changes even in ſerene weather; and an
equal difference would take place in aſcending one
thouſand four hundred feet from the ſurface; yet in
neither of theſe caſes are we conſcious of any alteration in the force of ſound. Much undoubtedly muſt
be aſcribed to the ſtillneſs and obſcurity of the night,
when the exerciſe of the other ſenſes is in a manner
ſuſpended, and that of hearing engages almoſt the
whole attention. During the meridian heats, alſo,
various noiſes, the warble of birds, the hum of
inſects, and the chearing calls of rural labour, at once
aſſail the ear, and render that organ leſs ſuſceptible to

other

other impreſſions.   But there is another cauſe which
ſeems generally to have been over-looked, though I
am convinced that it contributes moſt of all to the
effect.

The vivacity of the intimation given by the ſenſes
depends not ſo much upon the force of the impreſſion as upon its ſimplicity and diſtinctneſs.   Hence,
in high winds, ſound is heard at only a ſhort diſtance,
becauſe the aerial undulations are then diſturbed and
confounded.   In calm weather, too, when the ſky is
clear, and the ſun-beams act fiercely on the ſurface
of the earth, the heated air continually aſcends, and
this inteſtine motion deranges the undulations in their
horizontal progreſs.   After the cloſe of the day,
an equilibrium of temperature is again eſtabliſhed
between the higher and lower ſtrata of the atmoſphere,
which, during the tranquillity of the night, receives
and propagates diſtinctly every impreſſion.

Vol. I. Page 20. Line 8.   [*Specific Gravity of Birds.*]

THE lightneſs of the feathers and the hollowneſs
of the bones of birds have generally been aſ-
ſigned as the chief cauſes of the rapidity of their
flight.   And ſome naturaliſts, giving reins to their
imagination, have alleged, that, as the cavities are
filled with a ſort of inflammable gas, theſe animals are
buoyed up in the atmoſphere like balloons.   Such
reaſonings are not ſuperficial merely, they are abſurd.
The ſpecific gravity of quadrupeds is hardly inferior
to that of water, and therefore about nine hundred
times greater than that of air; and admitting that in
birds an equal quantity of matter occupies a triple
ſpace, which is ſurely an ample conceſſion, they would
                                                    loſe

lofe only the three-hundredth part of their weight at the furface of the earth, and ftill lefs in the fuperior regions of the atmofphere. Will any perfon infift on the efficacy of fo trifling a caufe? Nay, the diminifhed gravity of birds, far from affifting their flight, produces the oppofite effect in a very high degree; for the refiftance of the air, the great impediment to their motion, is proportioned, when other circumftances are alike, to the extent of furface which they prefent. This obfervation is remarkably exemplified in the Bird of Paradife, which is clothed with fuch a profufion of feathers, that it cannot face the gentleft gale, but is carried involuntarily into the ftream.

The obftruction which birds encounter in their flight, is much more confiderable than might at firft be apprehended: and this pofition is evinced by a very obvious fact:—Moft fpecies fly apparently with equal eafe, whether before or againft a moderate wind, and therefore the ftroke of the blaft is greatly inferior to the ordinary refiftance experienced in their paffage through the air.

The rapid flight of birds refults wholly from the prodigious power exerted by their large pectoral mufcles. This force may be refolved into two portions; the one employed in fupporting the birds in the air, the other in impelling it through that refifting medium. The former is conftant, and proportioned to the mafs; the latter depends on a variety of circumftances,—the quantity of furface, the fhape, the velocity, and the denfity of the furrounding element. The relative proportion of thefe two forces muft therefore vary extremely. In very large birds, their cumbrous weight can hardly be borne up, while their quantity of furface, which is comparatively fmall,

occafions

occasions not any considerable obstruction to their motion. It is thus that the Ostrich supports her body by means of her feet, and carries herself forward chiefly by the action of her wings. On the other hand, the surface is so great in proportion to the weight in very small birds, that almost their whole exertions are employed in overcoming the resistance of the air. If they intermit the strokes of their wings, the motion they have acquired is quickly extinguished: thence the sudden deflections which distinguish their flight. Of this, a remarkable instance is the Humming Bird, which, for its fluttering irregular progress, has been aptly compared to the humble bee.

Nothing contributes so much to facilitate the motion of Birds through the air, as the acuteness of the angle formed between their shoulders and their bill. For that reason, they extend their head, and endeavour to give their body as taper a shape as possible. Hence also, the birds which are most remarkable for their fleetness, have generally long necks ; such are most of the sea-fowl, which undoubtedly exercise their wings more than those of the land.

The resistance which a body suffers in its passage through a fluid is proportional to the square of the velocity. In slow motions, therefore, it is inconsiderable, but accumulates most astonishingly with an increase of celerity. Hence birds that differ widely in point of strength, fly pretty nearly with the same rapidity ; for it would require four times the force to give double the velocity, nine times to give the triple, and so forth. We likewise see the reason why the difference is not very great between the ordinary flight of a bird, and that wherein it exerts itself to the utmost.

In

In fimilar cafes the force neceffary to impel a bird through the air is proportioned to the denfity of that medium. It will fly therefore with moft eafe in the higher regions of the atmofphere; but this advantage is modified, and often over-balanced, by another cir-cumftance. The weight of the bird requires con-ftantly the fame force to fupport it; and this force, in the prefent inftance, can be produced only by the greater celerity of ftroke; a condition which is not always compatible with the ftructure of the animal. This inconvenience will be chiefly felt by the larger fpecies of birds, which, for that reafon, can never rife to any vaft height. The little tribes, on the contrary, are invited to foar far beyond the region of the clouds, where they glide with wonderful facility. What alone feems to fet bounds to their afcent, is the cold which prevails at thofe heights. And this influence is not fo great as might at firft be apprehended; for a ftream of rarefied air operates flowly in robbing a body of its heat, and therefore gives a weak fenfation of cold. It is extremely probable that fmall birds rife three or four miles into the air, where that fluid has only half the denfity that obtains at the furface. At that height the cold will indeed be fixty degrees; but its effects will be equivalent to a cold of only thirty degrees at the furface of the earth; which is lefs than the difference that often happens between the temperature of our fpring and that of our fummer. The care of provid-ing food commonly detains them indeed near the fur-face; but when they retire into other climates, they mount to the lofty regions of the atmofphere, and purfue their arduous journey far beyond the reach of human fight. No wonder then that the migrations of the fmall birds fhould have given occafion to fo much difputation; while thofe of the larger fpecies, fuch as the Goofe, the Stork, and the Crane, have been univerfally admitted.

As a bird in flying is actuated by two forces, the one impelling it upwards, the other forwards; the ftroke of its wing muft be performed in an oblique direction, between the vertical and the horizontal; and it will be more inclined to the latter in proportion to the fmallnefs of the bird, and the fwiftnefs of its motion. This is manifeft in the cafe of Pigeons, which are fo noted for their rapid flight. The pofition of the tail alone might indeed determine the direction of a bird's track; but that expedient would be attended with an expence of force which Nature has employed with fuch frugality. In fhort, it is extremely probable, that from obferving the infertion of the wings, a phyfiologift could infer with tolerable accuracy, the ufual rate at which a bird flies.

Birds often feem to reft fufpended in the air; but the appearance is illufory, for the force required to fupport them is in every cafe the fame. Either they fuffer themfelves to fink gently on their expanded wings through a certain fpace, and then by a few lengthened ftrokes, recover their former ftation; or they maintain their place by the nimble and vigorous quivering of their pinions, which is frequently difcernible.

The tail of a bird has often been compared to the rudder of a fhip; but the analogy is incomplete; for the motion of a fhip is confined invariably to the fame plane, while that of a bird is performed in every poffible direction. The pofition of the tail affects only the angle of afcent or defcent; it is the inclination of the head which turns the courfe to the one fide or the other.

The

The remarkable hollownefs of the bones, particularly thofe of the wings, though it by no means contributes to the effect afcribed to it in the text, is fub-
fervient to feveral very ufeful purpofes. It adds
greatly to the ftrength of the bones ; and this prin-
ciple feems to have directed Nature in many of her
animal and vegetable productions. Were the ftalks,
for inftance, of tall flender plants compacted into a
folid form, they would be unable to refift the fmalleft
violence. As the cavities of the bones in birds communicate with the lungs, they muft confpire to form
and augment the voice. Analogy clearly leads to this
conclufion, fince the *antrum*, or fmall cavity near the
bottom of the frontal bone, at the origin of the noftrils,
is found by experience to affift the human voice. But
the moft effential ufe of the hollownefs of the bones is,
perhaps, to afford an ample furface for the infertion of
the powerful mufcles.

Feathers, like the fly in mechanics, ferve to equa-
lize the motion of birds, but at a great expence of
force. Their principal ufe, however, is certainly to
confine the animal warmth which is generated, or rather evolved, by the procefs of refpiration. Loofe
fpungy fubftances, fuch as cotton, hair, wool, and
particularly feathers, are flow conductors of heat,
and therefore admirably calculated for the purpofe of
clothing. The wafte of vital heat on the furface of
the body is occafioned by the fucceffive contact of
air, and proportioned to the quicknefs of the application. It is hence that a ftrong wind will, even in
temperate weather, affect us with fenfations of cold,
though we often feel very comfortable during hard
froft, when the air is ftill. Birds therefore, more than
any other animals, muft be expofed to this wafte of

heat ;

heat; and accordingly they are clothed with a thicker and richer garb. We may also remark, that small birds are for their size better feathered than large ones, as the surface which they expose to the cooling stream of air is proportionally greater. It is thus that Nature kindly suits her provisions to the wants of her creatures. Man alone is sent into the world naked and helpless, and perpetually urged by his necessities to the exercise and cultivation of his faculties. This view of the utility of physical evil is finely illustrated by the elegant Virgil.

> " ——————————— Pater ipse colendi
> Haud facilem esse viam voluit, primusque per artem
> Movit agros, *curis acuens mortalia corda:*
>   *   *   *   *
> Ille malum virus serpentibus addidit atris,
> Prædarique lupos jussit, pontumque moveri,
> Mellaque decussit foliis, ignemque removit,
> Et passim rivis currentia vina repressit:
> Ut varias usus meditando extunderet artes
> Paulatim, et sulcis frumenti quæreret herbam,
> Et silicis venis abstrusum excuderet ignem.

Geor. Lib. I. 129.

> The fire of gods and men, with loud decrees,
> Forbids our plenty to be bought with ease;
> And wills that mortal men, inur'd to toil,
> Should exercise, with pains, the grudging soil.
>   *   *   *   *
> Jove added venom to the viper's brood,
> And swell'd with raging storms the peaceful flood:
> Commission'd hungry wolves t' infest the fold,
> And shook from oaken leaves the liquid gold.
> Remov'd from human reach the cheerful fire,
> And from the rivers bade the wine retire:
> That studious need might useful arts explore,
> From furrow'd fields to reap the foodful store;
> And force the veins of clashing flints t' expire
> The lurking seeds of their celestial fire.

Dryden.

END OF THE FIRST VOLUME.

Printed in the United States
By Bookmasters